AF477526

FORMAL SPECIFICATION AND VERIFICATION IN VLSI DESIGN

BRUCE S. DAVIE

FORMAL SPECIFICATION AND VERIFICATION IN VLSI DESIGN

EDINBURGH UNIVERSITY PRESS

© Bruce S. Davie 1990

Edinburgh University Press
22 George Square, Edinburgh

Printed in Great Britain
by Redwood Press Limited
Melksham, Wilts

British Library Cataloguing
in Publication Data

Davie, Bruce S.
Formal specification and verification
in VLSI design.
1. Computers. Design. Applications
of computer systems.
Description languages
I. Title   II. series
601.392

ISBN 0 7486 0159 7

CONTENTS

ACKNOWLEDGEMENTS

First and foremost I wish to thank George Milne who supervised my Ph.D. work which led to this book. His advice and encouragement, not to mention the provision of shelter, meals and beer above and beyond the call of duty, were invaluable. Nik Traub, Mauro Pezze, David Rees and Robin Milner also contributed through helpful discussions, while Shen-hong Zhu provided some enthusiastic technical support. Many thanks are also due to Joel Gannet and Mary Sheeran for their perseverance in reading and reviewing the book, and to Richard Clayton, who was an endless source of wisdom in matters of typesetting.

The research for my Ph.D. was made financially possible by a scholarship from the Association of Commonwealth Universities and the British Council, with additional assistance from Texas Instruments Ltd. Bell Communications Research generously allowed me the time to turn the thesis into book form.

Final thanks go to my wife Jody who, even when on the other side of the Atlantic, was a continual source of moral support.

ONE

INTRODUCTION

1.1   Rationale

To begin, it is worth considering why formal specification and verification techniques are important in the design of Very Large Scale Integrated circuits (VLSI). Formal verification is a particular approach to validation, and its advantages over other approaches will be the subject of later chapters. For the moment, the more general question of why validation is important will be addressed.

It is hard to overstate the importance of validating a large, complex design. It is clearly important that a chip functions as required in any application for which it will be used; i.e., it must conform to some specification. One way to establish this is to test the chip extensively after it has been fully designed and fabricated. However, the high cost of fabricating VLSI chips, the time taken to put a design on silicon (or some other material), and the difficulties that may be encountered in testing a fabricated chip due to unobservable internal behaviour make this approach extremely unattractive. Thus the vast majority of designs are simulated in some way prior to fabrication in the hope that any design errors will be detected before incurring the high costs, in both time and money, of fabrication. A wide range of simulation techniques is available and is discussed below.

To demonstrate the importance of formal specification techniques, it is necessary first to consider the integration of validation into the design process. It is common for VLSI design to be performed hierarchically in order to reduce the complex problem of design into subproblems of more manageable size. The levels in the design hierarchy are often called levels of abstraction, as devices become more abstract at the higher levels. There is some debate, however, in the VLSI field as to whether hierarchical design should be approached 'top-down' (starting from an abstract, high-level specification and moving towards a more concrete description of the circuit) or 'bottom-up' (constructing successively more complex devices from simpler devices that already exist). In general a mixture of the two techniques may be most appropriate. This issue will be discussed in later chapters.

Although it is possible to perform validation as an integral part of the hierarchical design process, it is generally more common for validation to be treated as a post-design exercise. The most common commercial simulators can model the behaviour of only very 'low-level' components such as transistors, capacitors, etc. at the circuit level [Nagel 75] or switch level [Bryant 81, Bryant 84, Terman 83] or gates, inverters etc. at the logic level [Hayes 86] thus requiring that the design be completed to that level before it can be simulated. Because such simulators can only model the behaviour of devices at a low level of abstraction, any hierarchy that was present in the design is lost at this stage. The problem with post-design validation for VLSI is that the amount of work required to take a design to the stage where it can be simulated is extremely large; often it may be measured in man-years. Thus the cost of finding a design error at such a late stage can be high as it may require the redesign of large sections of the chip. Ideally one would like to be able to detect errors at the earliest opportunity to minimise the amount of wasted design effort. This capability is offered by incorporating the validation task into the hierarchical design process.

1.1.1 An Integrated Design and Validation Methodology

For the purposes of the following discussion it will be assumed that design is to take place top-down, although, as suggested above, it is common for design to be a mixture of top-down and bottom-up techniques. For the moment this is an acceptable simplification to make. It will be made clear later how some of the ideas are also applicable to bottom-up design.

The validation of a design is the process of establishing that it conforms to a specification. Thus the essential first step of a hierarchical design methodology that aims to integrate validation must be to create a specification of the required behaviour of the system. (The precise meaning of 'behaviour' will be discussed later. At this stage a definition along the lines of 'what the circuit does', as distinct from how it does it or how it is constructed, will suffice.) It is common for top-level specifications to be fairly informal, perhaps using a mixture of representational forms such as English language sentences, timing diagrams, tables of timing data, etc. The aim here is to show why such specifications should be formal. Lack of ambiguity is one major consideration, but the role of formal specifications in hierarchical design and validation, which will be explained below, is the most important.

At this point it is important to define some terms. (A complete set of definitions appears in Appendix A.) Throughout this book, a specification will be assumed to be a description of a device's behaviour that gives no direct information as to how the device might be constructed — it could also be called a 'black-box' description. This corresponds to the fact that a pictorial representation of the circuit at this stage would just be an empty box, containing no information about its internal structure. This first specification has already been referred to as 'top-level'; this implies that the specification represents the highest (most abstract) level in the design hierarchy. Note that at any level in the hierarchy there may be both structural and behavioural (and possibly also geometrical) descriptions of the design. Thus a behavioural description is not intrinsically 'higher-level' than a structural one. Rather than referring to behaviour, structure and geometry as different *levels* of description, it is better to consider them as different descriptive *domains*.

In order to move down to the second level of hierarchy the designer must partition the top-level 'box' into smaller parts and ascribe a behaviour (either formally or informally stated) to each of these parts. In the traditional approach to hierarchical design these parts would each become a new black box. Each of them would then be treated as a system in its own right, which would be further partitioned, thus creating more and more levels in the hierarchy. At some point a bottom level is reached. For example, in VLSI design the bottom level is usually the level at which the circuit can be described in terms of interconnected wires, transistors and capacitors, while in a standard cell design system the bottom level might consist of gates, counters, latches etc. When all parts of the system have been designed down to this level, the design is considered complete.

It has already been mentioned that if validation is left until the design has progressed to the bottom level the cost of a design error may be very high. Thus it is proposed that validation should take place between every pair of adjacent levels in the design hierarchy. This is illustrated in Figure 1.1. In this book, a lower level in the hierarchy will be referred to as an implementation of the higher level. The behaviour of the higher level, whether it is the top one or not, is a specification. Thus the validation task consists of trying to show that the behaviour of an implementation satisfies a specification.

In order to do this it is *essential* that the boxes which constitute the design at every level are formally specified in some language. The language that is used to write the behavioural description must either be suitable for input to a simulator or support mathematical reasoning about the behaviour of the constructed device. In this way the behaviour of an implementation can be established and then compared with the specification behaviour. If at any point the implementation is found to be incorrect, i.e. does not satisfy the specification, then the designer must return to the preceding level in the hierarchy and repeat the process of partition-

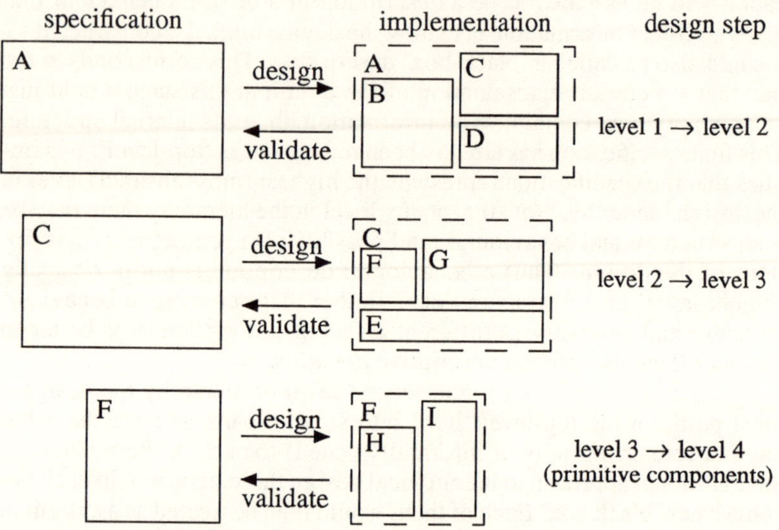

Figure 1.1: Validation in hierarchical design

ing the current box into parts and specifying their behaviours. Note that the amount of work wasted here is minimised, because the error can be detected before the design has been carried out down to the lowest level in the hierarchy.

Thus, the argument for formal specifications can be summarised as follows: validation should be performed between all levels in the design hierarchy to minimise the cost of errors, but such validation is only possible if formal, behavioural specifications are written at every level of the hierarchy.

By far the most common approach to formal specification is the use of hardware description languages (HDLs). These languages have been the subject of much attention in recent years and the ability to use the languages to describe behaviour has been of particular importance. Much of this book will deal with the development and use of hardware description languages with behavioural description capabilities.

1.2   Subtasks in Hierarchical Design and Validation

In this Section the methodology described above is examined more closely. The step of moving from one level in the design hierarchy down to the next level can be seen to consist of a number of subtasks. These tasks are illustrated by Figure 1.2. This diagram represents an attempt to imple-

ment a box 'A' by partitioning it into three boxes 'B', 'C' and 'D'. Rect-
angles are used to represent structural entities while ovals represent the
behavioural descriptions corresponding to these structures. The specification of 'A' must be written first if 'A' represents the whole system being
designed. If however 'A' is just one of several boxes at a lower level in a
design hierarchy its specification will already exist, having been written
in the 'describe' phase at an earlier step. Thus there is no real difference
between the two tasks of specifying and describing other than the fact
that the former applies to the higher level of abstraction and the latter
applies to the lower level.

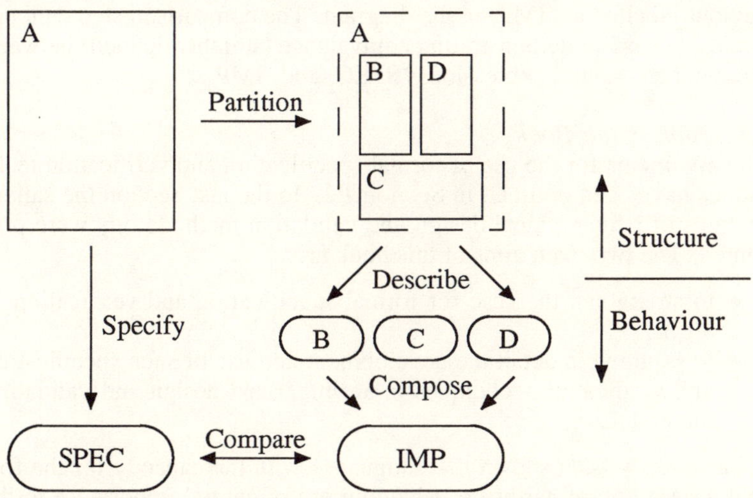

Figure 1.2: Subtasks of a design/validation step

The 'partition' task is a purely structural one which involves splitting
the higher level box into smaller boxes and describing their interconnections. It may be considered rather artificial to separate this from the
description task, since a designer will invariably have an informal idea
of the behaviour of each lower level box when he performs the partitioning. If, for example, box 'C' was actually called 'register' then this
would imply a certain behaviour for that box involving the storage of
data, clocking and so forth. The reason for showing description as a
separate task is that the *formal* description of the behaviour of each of
the boxes 'B', 'C' and 'D' must follow the partitioning task. These two
subtasks make up the 'design' step shown in Figure 1.1.
The validation step is also made up of two parts, the subtasks labelled

'compose' and 'compare' in Figure 1.2. Precisely what these tasks entail depends on whether the validation technique used is simulation or formal, mathematical proof, which will hereafter be called verification. (It should be noted that the term 'validation' will be used to include both simulation and verification.) If simulation is chosen, the composition may be considered to be performed by the simulator, whose task is to reproduce the behaviour of the circuit which consists of boxes 'B', 'C' and 'D' wired together in a certain way. This behaviour would then be compared, either automatically or by inspection, with the behaviour specified for the box currently being designed, 'A'. If formal verification is to be used, then composition is performed mathematically to establish the behaviour labelled as 'IMP' in the diagram. The comparison step then becomes the task of demonstrating equivalence (suitably defined) between the two behavioural expressions 'SPEC' and 'IMP'.

1.3   Aims of this Book

The arguments for the use of formal specification and verification techniques have been outlined in Section 1.1. In the last Section the salient features of a hierarchical design and validation methodology were presented. The two main aims of this book are:

- to strengthen the case for formal specification and verification in VLSI;
- to examine in detail the development and use of such specification and verification techniques in an integrated design and validation methodology.

It has already been shown that languages with the capacity for the formal description of hardware behaviour are of central importance to the proposed approach. The nature and use of such languages will therefore be a main theme of the following discussion. In seeking to achieve the above aims, the advantages of a language-based approach to design will be demonstrated, and the language features that are required will be examined. It will be shown how a suitably defined language may support an integrated, hierarchical approach to design and validation such as that described above.

Because of the importance of languages for the description of hardware behaviour, the body of the book begins with an introduction to the field of behavioural hardware description. Several languages with behavioural description capabilities will be described along with examples of their use. The CIRCAL descriptive framework [Milne 83a, Milne 84, Milne 85a] will be used as the main example to illustrate the ideas of the methodology, its subtasks, and the role which a language may play in these subtasks.

CIRCAL differs in an important way from most other hardware description languages: it was developed to support formal reasoning about the behaviour of hardware systems rather than simply for the description of hardware. It is related to Milner's CCS [Milner 80, Milner 83], as both formalisms grew out of work by Milne and Milner on the modelling of concurrent systems [Milne 79]. The fundamentals of CIRCAL and its use for describing hardware will be presented in Chapter 2.

The next three chapters deal with the three main tasks of the methodology described above: specification, design and validation. In Chapter 3 the problems of writing accurate and concise specifications will be addressed. The importance of language features to assist in this task, and the need for specification techniques coupled with the language, will be demonstrated. Various enhancements of the CIRCAL language to increase its usefulness for specification will be proposed.

In Chapter 4 the design task, comprising the subtasks of partitioning and description, will be discussed. Ways in which a language may assist design will be presented, along with design techniques to assist validation. Fully manual, transformational, and automated approaches to design will be discussed along with the requirements that each of these approaches place on the design language.

The final task to be discussed is validation. The two approaches of simulation and formal verification will be presented and their respective merits compared. As with design, these different approaches place different requirements on the description language and its associated specification techniques. Some of the techniques proposed in Chapter 3 will be justified and some further techniques will be developed.

Following this treatment of the individual subtasks, the ideas that have been presented will be illustrated by applying them to a reasonably large example. This will consist of the specification of a simple computer, its design to a lower (but still quite abstract) level of hierarchy, and the formal verification of the design step. Chapter 6 will demonstrate the effectiveness of the techniques for specification, design and verification that have been proposed in previous chapters.

All of the subtasks may be affected by the use of constraints, a term that has recently received increasing attention. In Chapter 7, constraints are defined and the ways in which they may affect and provide assistance in each of the previously discussed tasks are described.

Recurring themes of this book will be the close relationship between the tasks of specification, design and validation and the importance of languages in the execution of these tasks. Given the central role of languages, a logical first step is to investigate what it is that languages must describe and to examine some existing languages. This is done in the following Chapter.

TWO

HARDWARE DESCRIPTION LANGUAGES

In the preceding Chapter, the advantages of performing validation between each adjacent pair of levels in a design hierarchy as an integral part of the design process were demonstrated. In order to carry out such an approach to design and validation, there must be a mechanism for comparing the behaviours of design components at different levels of description in the hierarchy. This comparison is made between a behavioural description of a 'black box' at the higher level of abstraction (the *specification*) and a behaviour (the *implementation* behaviour) which is constructed using knowledge of the interconnection of parts and their respective behaviours at the lower level. It is therefore necessary to be able to describe behaviour at any level of abstraction; it is also necessary to describe the interconnection of parts, i.e. the *structure* of the circuit. Structure and behaviour are referred to as descriptive domains; there is a third domain, geometry, which is usually described in special-purpose languages such as CIF [Mead 80] and EDIF [Carlstedt 86].

It is with the aim of facilitating the rigorous description of circuit behaviour and structure that many hardware description languages (HDLs) have been developed in recent years. The motivations for producing these languages have been many and varied and this is reflected in the diversity of languages which have arisen. It has already been seen that behavioural description is essential for hierarchical validation; it has also been required for documentation purposes [Lipsett 86], as a medium for the assistance of designers [Morison 85] (analogous to the use of circuit diagrams to help designers visualise structures), as input to design automation tools such as Macpitts [Siskind 82], and to provide a medium for the transfer of design information between different design tools (for example, [SCS 87a]). In the following Section a number of the recently developed languages will be examined and their salient features will be described and compared. Particular attention will be given to CIRCAL, which will be used as the primary example of a language for the integration of specification, design and validation. Before focusing attention on any particular languages, however, it is appropriate to consider more closely the issue of behavioural description.

2.1 Describing Behaviour

Before discussing how to describe behaviour and thus how one might choose to define a language to do this, the meaning of behaviour in the context of VLSI or circuit design must be established. The way in which behaviour is perceived will have a great effect on the features required for its description. It is helpful to consider some common pieces of hardware to determine possible meanings of the term 'behaviour'. The behaviour of a piece of combinational logic, for example, might be characterised by the logical function which it computes. This could be represented as a mathematical function whose arguments are the values on the input ports of the piece of logic and whose result is a set of values on the output ports. In a very general sense, then, behaviour might be considered as the dependency of certain values (outputs) on certain other values (inputs). Key concepts to note here are *values* and *ports*. Another piece of information which might be important to the behaviour of the piece of combinational logic is the length of time it takes to compute the function. This may be measured as the amount of time between a new set of values being presented on the inputs and the new computed value appearing on the output. This introduces two more concepts of behaviour: the idea of *changes* on ports (i.e. the arrival of 'new' values) and that of the passage of time. Time may be measured qualitatively (certain events happen after other events) or quantitatively.

Combinational logic represents a very simple class of circuits, in which the output is a time-independent function of the inputs at a single instant in time, requiring no knowledge of their past history. The other main class of devices consists of those whose behaviour does depend on past history, i.e. *sequential* devices. Thus one more behavioural concept, *state*, which is a function of past events and determines future behaviour, may be added to the list. These concepts — values on ports and changes of these values, the interdependency of changes and the passage of time between them, and state — are common to devices of widely varying complexity, from microprocessors to pass transistors. Describing the function of a microprocessor could be a very laborious task, but the idea of port values changing over time in response to other changing port values is still applicable.

While hardware description languages have been used to describe behaviour, structure and geometry, it is really only in the behavioural domain that they are recognised as the preferred way to describe design characteristics. It is certainly easier for a human to examine a colour plot of a piece of a VLSI design than to read the CIF file or a similar geometric description. Similarly, the majority of designers find that structure is much more naturally represented by circuit diagrams than by languages, although it may sometimes be necessary to use a linguistic description of structure as the input to a design tool (e.g. Chipsmith [Lattice 85]). For behaviour, however, it is difficult to come up with a reasonable alternative to a language, with attempts to represent behaviour diagrammatically being generally unnatural or limited in the range of devices which can satisfactorily be described.

Having established to some extent the meaning of behaviour it is now possible to discuss the requirements of a language for its description. It has been indicated that the central concepts of behaviour are common to a wide range of devices. Furthermore, in the methodology proposed it is essential that behaviour can be described in such a way that comparison between behaviours at adjacent levels in the hierarchy can be made. This suggests that a language should provide capabilities for the description of behaviour at a wide range of levels of abstraction. Most languages are not designed for accurate behavioural description below the gate level, since at this point some of the modelling assumptions cease to be valid, e.g. the concept of a signal's strength, not just its value, becomes important, and it may be necessary to take analog effects into account. Relatively few languages have facilities to cope at this level of description. At the other end of the spectrum the main problem is coping with descriptive complexity, and thus a language must have sufficiently powerful constructs to deal with this.

A key requirement for a hardware description language is some means of representing time. It is this feature which most clearly distinguishes HDLs from programming languages, the vast majority of which have no facility for the description of time. As will be discussed in the following Section when some particular languages are examined, there are a number of different ways in which time can be dealt with and these offer varying degrees of generality and ease of use. The type of timing phenomena which a designer may wish to describe depends heavily on the level of abstraction at which he is working. At the microprocessor level, for example, a designer may be interested simply in what happens in the period of execution of a single instruction, whereas at the gate level he may be interested in the time taken for a change of value to propagate across a device, or even the time taken for the value on a port to move from 'false' through 'undefined' to 'true'. If a language is to be used at all levels of abstraction, it must have sufficient generality to cope with this.

The concept of changes of values is also critical to hardware description. It is a familiar concept to circuit designers, for example in the description of an edge-triggered latch. Thus a language should in some way allow the writing of descriptions which refer to value changes. As might be expected, there are also a number of different approaches to this problem. These will be discussed in the following sections.

In summary, the most basic requirement for a behavioural description language to be used in the proposed methodology is that it be able to describe any type of hardware over a range of levels of abstraction. With this as a starting point, the key features of a hardware description language should be:

- a wide spectrum of description, ranging from gates (or lower) up to microprocessors or even more complex systems;
- a means of describing the interdependency of values on ports;
- the facility to refer to changes on ports;
- facilities for the description of timing, with sufficient generality to allow timing description at all levels of abstraction supported by the language.

In the following Section a number of languages will be presented which, although not necessarily designed with the express aim of supporting integrated, hierarchical design and validation, satisfy the above requirements to some extent.

2.2   A Selection of Languages

The number of Hardware Description Languages which have appeared over the last few years is very large, as some recent surveys indicate [Camurati 87, Nash84]. This Section deals with just a few of them but aims to provide some insight into both the features which are common to most languages and the considerable diversity of languages which exist. All of the chosen languages fulfil the basic requirements outlined above, yet they have significant differences in terms of the type of environment in which they were developed (e.g. academic or commercial), the main application(s) for which they were intended and, therefore, the features beyond the basic ones which were considered sufficiently important to be included in each language. They also differ substantially in the way in which they provide these features.

The languages presented below use widely varying notation. In general, this is explained as necessary by way of examples. Since the notation of both predicate logic and CIRCAL will be used repeatedly, summaries of both sets of notation are provided in Appendix B.

2.2.1   VHDL — The VHSIC Hardware Description
Language

In addition to the selection criteria listed above, this language is worthy of discussion simply because of its importance to industry and the number of man-years that have been invested in its development. Associated with the Very High Speed Integrated Circuit (VHSIC) programme, the main motivation behind its development was to provide a standard HDL for all hardware design projects within that programme [Dewey 86]. It was subsequently adopted as an industry standard by the IEEE [IEEE 87]. The magnitude of the VHSIC programme is such that any HDL which aims to serve it must be an extremely general one. Providing a medium for accurate specification and documentation of contracts was a high priority for the language. The language was designed by a large committee and underwent many reviews in an attempt to ensure that all the features that might possibly be required were included. The necessary features have been listed as [Aylor 86]:

- ability to describe a wide range of hardware;
- facilities for design management;
- timing description capabilities;
- architectural description capabilities;
- ability to describe a design's interface;
- ability to describe a design's environment;
- language extensibility;
- an implementation of the language which allows many tools to be driven by it;
- language semantics which are independent of any particular implementation;
- 'user-friendliness';
- programming language-like appearance for procedural parts of the language;
- ease of extraction and insertion of documentation data.

It is clear that these requirements go far beyond those proposed in the preceding Section. This is partly because the language is required to do more than support a hierarchical design and validation methodology, and partly because the developers of the language wanted it to be not just adequate but also easy to use. The result of this is that the language has many features which are of no great relevance in this discussion. The primary concern here is the description of behaviour in such a way that hierarchical design and validation may take place. As structural description is also necessary in order to establish implementation behaviours and thus perform validation, this will also be examined.

All devices described in VHDL must have an *entity* declaration. This describes the interface of the device, associating a list of typed ports with the named part. No information about the device's internal structure or behaviour is given. The following example shows an entity declaration for a RAM.

Example

```
entity RAM is
      port(
         DATA:      inout tristate_vector;
         ADDR:      in bit_vector;
         CS1,CS2:   in bit;
         RW:        in bit);
end RAM ;
```

Explanation

The capitalised words in the port declaration are the port names, and they are followed by pairs of words which describe the direction of the port (`inout` implies a bidirectional port) and its type. Note that vector types are used to enable the data and address ports, each of which physically consists of a number of wires, to be described as single ports.

Structural Description

In describing the structure of a device, its component parts must be declared and instantiated within an *architecture* declaration. The declarations of the components should match their own entity declarations. In order to specify the way in which these components are wired up, their ports are renamed in such a way that all ports which are intended to be connected to each other are given the same name. (A similar convention is used in many other languages, e.g. MODEL [Lattice 85], CIRCAL, etc.) The names of the wires used for interconnections within the device must be declared as signals. Because all ports have a type, type-checking can provide a rudimentary check for design errors at this stage. The ports which will be connected to the interface of the whole device are renamed to match the interface ports to which they will connect. The renaming operation is done simply by replacing names in the ordered lists of ports which are introduced by the keywords `port map` within each component instantiation statement. This is illustrated in the example presented below.

Example

Figure 2.1 shows the structure of a full adder, constructed from two half adders and an OR gate. The following is a possible description of the structure:

```
architecture Structure_Desc of Full_Adder is
    component Half_Adder
        port(I1,I2: in bit;
             Carry: out bit;
             Sum: out bit);
    component Or_Gate
        port(I1,I2: in bit; O: out bit);
    signal a,b,c: bit;
  begin
    H1: Half_Adder port map (X,Y,a,b);
    H2: Half_Adder port map (b,Cin,c,Sum);
    O1: Or_Gate port map (a,c,Cout);
  end Structure_Desc;
```

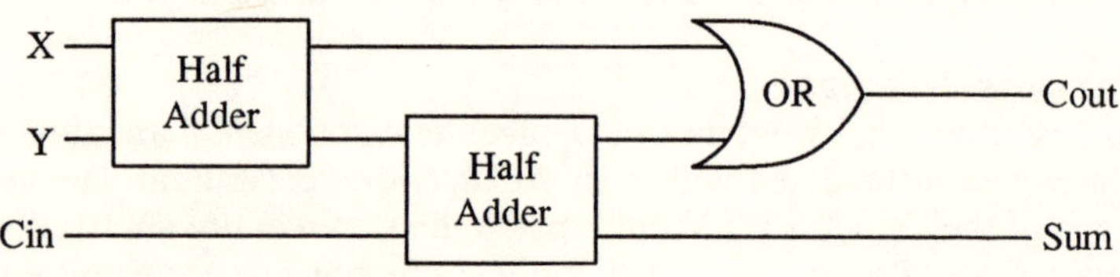

Figure 2.1: Structure of Full Adder

Explanation

The instantiations appear between the keywords `begin` and `end`. H1 and H2 are two instances of a component called Half_Adder, which will be defined in some way in a library or by the designer. The ordering of port and signal names in the port map list for H1 implies that its inputs are to be connected to ports X and Y, its carry output to internal signal a, and its sum output to signal b. In this way, all the connection information which is contained in Figure 2.1 is conveyed by this description in the language. Note that there is considerable redundancy in the description, as the declarations of `Half_Adder` and `Or_Gate` convey no information that would not be contained in the definitions of those parts. This probably reflects the aim of the language designers to provide a language suitable for documentation purposes.

Behavioural Description

There are several ways of representing behaviour in VHDL; these are explained in detail in the user's manual [IEEE 87]. The most general is the *process* statement, which can appear in an architectural body. Many common programming language constructs are provided for the description of algorithms and functions. The principal timing construct is a delay specified in physical units (e.g nanoseconds) using the keyword after. The following example, a behaviour of the RAM chip whose interface was defined above, illustrates the basic principles.

Example

```
architecture Behaviour_Desc of RAM is
  process(CS1,CS2,DATA,ADDR,RW)
    type matrix_item is
          tristate_vector (7 downto 0);
    type matrix is
          array (0 to 1023) of matrix_item;
    variable memory: static matrix;
  begin
    if CS1 and CS2 then
        case RW is
          when '0' =>
                DATA <= memory(IntVal(ADDR))
                          after 70ns;
          when '1' =>
                memory(IntVal(ADDR)) := DATA;
        end case;
      else
        DATA <= "ZZZZZZZZ" after 55ns;
    end if;
  end process;
end Behaviour_Desc;
```

Explanation
Most of this description looks just like a conventional programming language. A few symbols require explanation. The right arrow '=>' introduces each possible action for the case statement. Delayless assignment is represented by ':=' and delayed assignment by the left arrow '<='. The magnitude of the delay is specified using the after construct. Its interpretation in a statement like

```
DATA <= memory(IntVal(ADDR)) after 70ns;
```

is as follows: if any of the variables on which this statement depends (i.e. CS1, CS2 or RW) change, then a new value must be calculated for DATA. This new value will appear on the DATA port 70 nanoseconds after the change that caused it. Thus the underlying model of behaviour used is event-based, relying on the concept of changes on ports for the description of timing properties.

In order to make explicit reference to changes of value (to describe an edge-triggered device, for example) it is necessary to use a pre-defined attribute, 'STABLE'. *Attributes* are a VHDL feature which allow virtually any sort of information to be associated with a description. A user may define new attributes as they are needed, to drive new tools for example. This particular attribute is built in to the language and simply allows a signal to be tested for changes of value. Its use is illustrated in the following example, an edge-triggered D-latch.

Example

```
entity D_Latch is
      port(
          DATA, CLOCK: in bit;
          Q:               out bit);
end D_Latch;

architecture Edge_Behaviour of D_latch is
      process(DATA,CLOCK)
      begin
          if CLOCK and not CLOCK'STABLE then
             Q <= DATA after 15 ns;
          end if
      end process
end Edge_Behaviour;
```

The above description specifies that the value on the input DATA will be transferred to the output Q if a rising edge occurs on the CLOCK, with a delay of 15 nanoseconds.

Comments

From the brief introduction to the language given above, it can be seen that it exhibits the essential features which were proposed for behavioural description. The language has many more features than those illustrated here, a consequence of its designers' aim to create a language suitable for a wide variety of applications. The description of functional dependencies between the values on various ports is achieved with the use of common programming language constructs such as conditional statements and logical operators. Timing characteristics can also be described. Reference to changes of ports values is usually implicit, relying on the underlying event-based model of the language. With a small amount of extra effort, explicit reference to value changes can be made if required. However, the way in which this is done seems to suggest that the ability to refer to changes was added as an afterthought, rather than being considered a central feature of the language.

The provision of high-level programming constructs, such as the ability to define datatypes, contributes additional power to the language. For example, the ability to consider a bundle of wires as either separate wires, a vector, or a single integer greatly simplifies behavioural descriptions, especially for complicated devices at high levels of abstraction. The ability to define types may also be important at lower levels of abstraction, for example when describing tri-state devices. In general, the language is highly extensible, making it suitable for a wide range of applications and tools.

Subjectively, descriptions in this language seem to be highly readable (again, a result of the use of programming-language-like syntax) but rather verbose. Both these characteristics probably result from the desire of the language designers to provide a language for documentation purposes.

A final comment about this language regards the type of validation which it is intended to support. It is fully intended to be used as an input language for multi-level simulators, but it does not have the simple, well-defined semantics which would make it suitable for formal reasoning about behaviour and thus for verification. This is a common feature of HDLs: that the syntax is defined first, then attempts are made to attach meaning to the syntax. This contrasts with languages such as CIRCAL and HOL [Gordon 85], which begin with a framework supporting formal reasoning and then attempt to apply this to the description of hardware. These different approaches will be discussed in Sections 2.3 and 2.2.4 respectively.

2.2.2 LTS

A language that provides a sharp contrast to VHDL is LTS (Layout and Timing for Structures) [Babiker 83, Babiker 85]. The language was designed by a fairly small team with the aim of supporting a design transformation methodology [Milne 86b], depending on correctness-preserving transformations such as those described in Section 4.2. Like VHDL, it uses a programming language-like syntax to enable the description of dependencies of outputs on inputs, although in this case the language is a functional one, similar to ML [Harper 86]. This leads to a language which looks rather different from VHDL.

Structural Description

The functional style of LTS leads to a straightforward way of describing structure. Since all devices are described by functions that map inputs to outputs, the internal structure of a device can be represented by the appropriate composition of the functions which represent its component parts. The full adder of Figure 2.1 is again used to illustrate the use of the language.

Example

```
fulladder(X,Y,Cin) = (Sum,Cout)
    where
        (a,b) = halfadder(X,Y),
        (c,Sum) = halfadder(b,Cin),
        Cout = or(a,c)
    end
```

Explanation

This device description is very similar to a definition of a ML function, where the arguments of each function represent its inputs and the returned results correspond to outputs. Internal signals are represented by local value bindings, as in

```
(a,b) = halfadder(X,Y)
```

which labels the outputs of one of the half adders as a and b.

This type of description is intuitive and fairly succinct. It does not convey quite as much information as the equivalent description in VHDL, as the type information for the ports is not stated explicitly. Since the type of a port is essentially a behavioural rather than a structural feature, this is not unreasonable. Type checking can still be performed, as the type of any port can be inferred from the behavioural definition of the component to which it is attached. The description also contains none of the other redundant information that was present in the VHDL description of the same device.

An unusual feature of LTS is that it allows detailed geometrical information, such as the size of cells and the positioning of ports, to be attached to device descriptions. This facility is not, however, relevant to the discussion here, so these features will not be described in detail.

Behavioural Description

The basic entities for behavioural description in LTS are signals that map time to values. A behavioural description of a device is simply a function that maps input signals to output signals. In order to describe tim-

ing phenomena, a single built-in function is required. This function is
`last(x)`, which returns a signal whose value at any point in time equals
the value of the signal x at the previous instant in time. The concept of
a previous instant implies that time is viewed as progressing in discrete
steps rather than continuously. This approximation is frequently made in
hardware description and is generally acceptable as long as the steps are
sufficiently small for the application in question.

Functions defined in LTS need not always represent pieces of hardware. For example, it is useful to define more sophisticated timing functions in terms of the primitive `last`. Some examples of useful functions
are:

```
delay(n)(x) =
    if n is
      0 then x,
      as m:1..? then delay(m-1)(last(x))
    end

rise(clock) =
    and(clock,not(last(clock)))
```

Explanation
The first of these functions defines a delay of length n. If n equals zero,
then the output of the function simply equals the input; otherwise, the
input signal is delayed by one unit using `last()` and the resulting signal is delayed by n-1 units by calling `delay` recursively. The second
function is true if the clock has risen at the most recent instant. These
functions are used in the next example.

Example
The device described below is a rising-edge-triggered D-type latch.

```
dlatch(data,clock) = q
  where
    q = if (rise(clock)) is
        true then delay(5)(data),
        false then last(q)
    end
  end
```

Explanation
The behaviour of the latch is defined as a function whose arguments are
the input signals `data` and `clock` and which returns the output signal

q. The behaviour of the output is defined so that if there is a rising edge on `clock` then the output will take the value which was held by the `data` port 5 time units previously. In any other circumstance, the output retains the value it held at the previous instant in time.

LTS provides access to the powerful datatype definition facilities of ML. This facilitates the writing of comprehensible and consistent descriptions and is particularly useful at very abstract levels. It also supports the use of ranges and arrays to assist in the description of the regular structures, which are common in VLSI design.

Comments

In some ways LTS provides similar behavioural description facilities to those of VHDL. Whereas the latter relies on imperative-style processes to define behaviours, LTS uses a functional approach. Both languages provide a tool for tackling complexity at abstract levels of description with powerful type definition capabilities.

A main difference between VHDL and LTS is the treatment of timing. The discrete-time, backward-looking model of LTS relies on a single primitive function, `last(x)`, which can be used as a basis for the definition of any number of more complex timing functions. It is noteworthy that, in contrast to VHDL, the LTS model of behaviour makes no reference to changes of values on ports (although functions to detect them, such as `rise`, can be readily defined), instead permitting behavioural description in terms of a history of values.

One appealing characteristic of LTS is that it seems to have been designed with the aim of achieving the desired degree of descriptiveness with the minimum number of primitive constructs. This perhaps reflects the fact that it was designed for use with a restricted set of tools, rather than aiming to serve a very large community of users with widely varying requirements. The result is a fairly intuitive and easily comprehended language, which is nevertheless very powerful. This contrasts with VHDL, which provides a very large, or even unwieldy, set of constructs with little if any gain in descriptive power.

2.2.3   The M Language

Whereas VHDL was developed to support many different tools from many vendors, the M behavioral modelling language[SCS 87b] was developed to support the multi-level simulator of a single tool vendor. And while the behavioral part of VHDL bears a resemblance to the Ada programming language, M's behavioral part is closely based on the C language [Kernighan 88]. In M the user has full access to the function and datatype definition capabilities of C. The basic descriptive entity is a *module*, which may contain structural or behavioral information or a mixture of both. A module looks much like a C function definition, except that it is preceded by the keyword MODULE and the definition contains up to three distinct sections, each delimited by various keywords. The uses of these sections are explained below.

Structural Description

The structure of a module can be defined in terms of other modules using the BUILD keyword. The following example illustrates this and shows how a module is declared.

Example

```
MODULE fulladder()
{
  IN X, Y, Cin;
  OUT Sum, Cout;

  BUILD {
    INSTANCE(halfadder,ha1);
    INSTANCE(halfadder,ha2);
    INSTANCE(or,or1);
    NET(ha1.I1,X);
    NET(ha1.I2,Y);
    NET(ha1.Carry,or1.I1);
    NET(ha1.Sum,ha2.I1);
    NET(ha2.I2,Cin);
    NET(ha2.Carry,or1.I2);
    NET(ha2.Sum,Sum);
    NET(or1.O,Cout);
  }
}
```

Explanation

Following the declaration of the module name, its terminals are defined. A terminal has a direction (IN, OUT or INOUT) and a type, which may be built-in or user-defined. The default type is BIT, which has been used here. In the BUILD section, the component parts are declared; the INSTANCE statement associates a name with each instance of a module that is to be used, so that there are half adders called ha1 and ha2 and an OR gate called or1. Each NET statement connects two or more terminals together, the syntax for terminal names being inst_name.term_name where term_name is taken from the definition of the module that is to be instantiated. A `term_name` alone (e.g `Cout`) refers to the terminals of the module being defined.

Although not illustrated in the above example, modules may be parameterised; an OR gate, for example, might have a parameter representing the number of inputs. Such a parameter would be fixed when the module was instantiated with the `INSTANCE` statement.

Behavioural Description

Two sections are used to describe the behaviour of a module. The section begun by the keyword `SIMULATE` contains the bulk of the behaviour, and the `INITIALIZE` section is used to ensure that initial signal values are correctly set at the start of a simulation. This section is executed only when a simulation is commenced, whereas the `SIMULATE` section is executed every time the value on any `IN` or `INOUT` port changes. Both sections contain statements, written in a superset of C, that assign values to output ports or to internal state variables. The latter may be declared using the `MEMORY` keyword. The following example, the same D-latch described above, indicates the use of some of the behavioural description features of M.

Example

```
MODULE dlatch()
{
  IN data, clock;
  OUT q;

  INITIALIZE{
    q = LOW;
  }

  SIMULATE {
    if(RISE(clock)){
      q = {data@5};
    }
  }
}
```

Explanation

Following the declaration of the ports (there are no `MEMORY` variables in this example), the `INITIALIZE` section simply sets the output to a predictable value for the start of a simulation. The `SIMULATE` section is executed each time a new value is detected on either `data` or `clock`, but the value of q will only be changed after a rising edge has been detected on `clock`. The `RISE()` function is one of several edge-detecting functions provided. The expression `{data@5}` enforces a transport delay of 5 nanoseconds; it is also possible to specify inertial delays. (The difference between these two delay types is explained in Section 3.2.3.)

Comments

Once again, the behavioural part of the language is conceptually not all that different from VHDL, consisting of a programming language-like syntax enhanced by additional constructs to enable description of the unique features of hardware. As in VHDL, structure and behaviour may, if desired, be mixed in the description of a single component. The description of structure seems more cumbersome than in most languages, with each connection needing to be described in a separate line.

The fact that M is designed to drive one particular tool (a simulator) has two notable consequences: first, there are several features (not all described here) which are directly geared to control of simulation, such as the `INITIALIZE` section. Secondly, the language does not aim to be as general as VHDL, which is expected to be used as a standard documentation language. M therefore does not provide the same range of features for documentation or the same level of readability; this is offset, however, by a reduction in redundancy and verbosity.

As an aside, it is interesting to note that the M language provides a complete set of facilities for the detailed description of analog behaviour. In addition, the simulator that is driven by M can also simulate (in either digital or analog fashion) modules whose behaviour has been extracted (by other tools) from pieces of VLSI layout. When establishing the behaviour of a constructed device by simulation, it is acceptable for some of the component parts to be described digitally in M, others in an analog manner, while others may have their behaviours defined by the extracted netlist. Detailed examination of the description of analog behaviour is beyond the scope of this discussion, but inclusion of this feature in M is sufficiently unusual to warrant comment.

2.2.4 *Higher Order Logic*

Another approach to hardware description is to use an existing descriptive framework rather than creating a new one. This approach has been taken by several researchers using higher-order logic to describe and reason about hardware, and offers the advantage of access to the reasoning capabilities of that framework. A mechanised system for theorem-proving in higher-order logic called HOL [Gordon 85] has been developed. The approach to hardware description described below follows that of Gordon.

Before examining the details of this approach, it will be helpful to introduce some notation. The symbols that will be used here are presented below. P and Q are predicates, x is a parameter.

$\equiv$	*is equivalent to*
$\neg P$	NOT P
$P \wedge Q$	P AND Q
$P \vee Q$	P OR Q
$\forall x.P$	*P is true for* all x
$\exists x.P$	*P is true for* some x
$P \supset Q$	*P implies Q*

For more complete information on predicate logic, the reader is referred to standard texts on the subject [Hatcher 82] or the numerous papers on hardware description using higher-order logic [Gordon 85, Gordon 86, Hanna 85].

The behaviour of a device is represented by a predicate whose arguments, representing the values on the ports of the device, are, in general, functions of time. This predicate defines the relationships which must exist between the current and past values on the ports of the device. As in LTS, time is assumed to proceed in discrete steps. This ability to write predicates whose arguments are functions of time is a key attraction of higher-order logic over first-order logic for the purpose of hardware description. For example, a NAND gate with a delay of one time step would be represented by the following predicate:

$$\text{NAND}(i_1, i_2, o) \equiv \forall t.o(t+1) = \neg(i_1(t) \wedge i_2(t))$$

The inputs to this device ($i_1(t)$ and $i_2(t)$) and the output ($o(t)$) are represented as functions of time and the predicate NAND is true when the output equals the nand of the inputs delayed by one time unit.

In general, the behaviour of pieces of hardware can be described as predicates making use of the standard notation of predicate logic such as conjunction, disjunction, negation, implication, etc. The following example shows how this notation can be used to describe the behaviour of a multiplier which takes an unspecified length of time to produce a result and raises the signal *done* when it has finished. The specification begins with the definition of some useful temporal predicates.

Example

$$\text{Stable}(t_1, t_2)(f) \equiv \forall t.t_1 \leq t \wedge t \leq t_2 \supset (f(t) = f(t_1))$$
$$\text{Next}(t_1, t_2)(f) \equiv t_1 \leq t_2 \wedge f(t_2) \wedge$$
$$(\forall t.t_1 \leq t \wedge t \leq t_2 \supset \neg f(t))$$

$$\text{Mult}(i_1, i_2, o, done) \quad \equiv \quad done(t_1) \wedge$$
$$\text{Next}(t_1, t_2)(done) \wedge$$
$$\text{Stable}(t_1, t_2)(i_1) \wedge$$
$$\text{Stable}(t_1, t_2)(i_2)$$
$$\supset$$
$$(o(t_2) = i_1(t_1) \times i_2(t_1))$$

Explanation

Stable is true if f is constant between times t_1 and t_2. Next is true if t_2 is the first time after t_1 at which $f(t_2)$ is true. Thus the meaning of the specification of the multiplier is that if the signal *done* is high at some time t_1 and subsequently at t_2, and the inputs i_1 and i_2 are stable during the interval, then the output at the end of the interval should equal the product of the inputs at the start.

Structural Description

The representation of structure in higher-order logic is straightforward. A device is constructed of several component parts. Each of these is described by a predicate which defines the relationship that must hold between the values on the ports of each part. The behaviour of the whole device must therefore be such that all of these predicates hold, i.e. it is the conjunction of the predicates of the component parts. This may be considered a structural description of the device: it tells how it is constructed. There may be some ports that are to be considered internal to the device. That is to say, the signals on these ports will not be communicated outside the device. In this case, the relationship between values on the external ports is all that matters, and this is represented by a predicate which is true for some values on the internal ports. The lack of a definite statement about the values on the internal ports is achieved by existential quantification. Thus, the structure of the full adder circuit of Figure 2.1 would be described as follows:

$$\text{Fulladder}(x, y, cin, sum, cout) \quad \equiv \quad \exists abc.\, \text{Halfadder}(x, y, a, b) \wedge$$
$$\text{Halfadder}(b, cin, c, sum) \wedge$$
$$\text{Or}(a, c, cout)$$

An important point to note here is that this 'structural description' is really more than that — it is a constructed predicate which fully describes the behaviour of the whole device in terms of the behaviour of its component parts. This distinguishes higher-order logic from the other descriptive frameworks seen previously, in which the behaviour of a constructed device could only be established by simulation. Because higher-order logic allows behaviours to be constructed in this way it is a suitable framework for formal verification. This issue will be addressed in Section 5.2.6.

Comments

It should be pointed out that the application of higher-order logic to hardware has been undertaken by several researchers and the exact approach taken varies. The methods illustrated above are modelled on those of Gordon, whose approach has been adopted by a number of others [Camilleri 86, Joyce 86]. Hanna and Daeche [Hanna 85] have taken a somewhat different approach, particularly suited to detailed modelling of devices at low levels of abstraction, based on the idea of partial waveform specifications. The model of timing adopted in Gordon's approach is essentially the same as that of LTS, i.e. a backward-looking discrete time model. In a similar manner to that seen in LTS, predicates which describe complicated timing phenomena can be defined as required using the primitive constructs of the language.

Two characteristics of higher-order logic distinguish it from the majority of hardware description languages. The first is that it is not a specially developed language, but a pre-existing framework which has been found to be suitable for the description of hardware behaviour and structure. A potential disadvantage of this approach is that the description of real hardware in this framework may be less intuitive than it is in the specially designed languages. However, the principal advantage of using logic to describe hardware is that the 'structural' operators, which describe the wiring up of parts to form a larger device, have a well-defined behavioural meaning, thus allowing the behaviour of the larger part to be established given the behaviours of its components. This is essential to formal verification, the subject of Section 5.2. Furthermore, by using the existing framework of logic, the rules of inference of that framework are immediately available for reasoning about hardware behaviour.

A final, palpable advantage of HOL is that a software system for the manipulation of HOL expressions exists [Gordon 85]. This system has enabled correctness proofs to be undertaken for large, complex designs such as the VIPER microprocessor [Cohn 87]. The use of HOL to perform proofs is further discussed in Section 5.2.6.

2.2.5 *Temporal Logic*

Another logical framework that has been applied to the description of hardware is temporal logic [Moszkowski 83, Moszkowski 85]. Whereas higher-order logic treats time as just another variable, temporal logic extends first-order logic with special operators for the representation of temporal properties. Concepts such as 'X is always true' and 'Y is sometimes false' can be represented using these operators, without reference to a variable t. The operators used to create expressions of this type are described below.

Basic concepts

Temporal logic formulas are referred to intervals, an interval consisting of a sequence of one or more states. The view of time is therefore discrete. The expression

$$\sigma \models p$$

means that the formula p is true in the interval σ. Omitting the σ implies that p is true in all intervals. The value of a variable in an interval is simply its value at the first state of the interval. If X is some formula then the expression $\Box X$ applied to a certain interval means that X is true at all states in the interval and $\Diamond X$ means X is true at some state in the interval. These are read 'always X' and 'sometimes X' respectively. The expression $\sigma \models X$, read as 'next X', is true if the interval has at least 2 states and X is true in the subinterval obtained by deleting the first state of σ.

Using just the three temporal operators described above and the conventional operators of predicate logic it is possible to define several more useful operators. For example, a unit delay may be defined as follows:

$$X \, del \, Y \equiv_{def} \Box(true \supset [(X = 0) \equiv (Y = 0)])$$

This predicate is true if X is equal to zero when Y is equal to zero in the next state. Given that X and Y may be only 0 or 1, then X must also equal 1 exactly when Y is 1 in the next state. The inclusion of the term *true* ensures that the predicate only applies to intervals with at least 2 states. Thus, the predicate is true if Y is equal to the signal X delayed by one time unit.

Application to Hardware

Structural description in temporal logic is similar to that seen in higher-order logic, since in both cases the operators of first order logic are used. For behavioural description, the basic operators of the logic plus a set of derived operators such as *del* make it possible to describe hardware devices in a way that takes account of their temporal behaviour. As a simple example, a 3 input NAND gate with unit delay is described below.

Example

$$nand3(A,B,C,Z) \equiv_{def} (\neg(A \wedge B \wedge C)) \, del \, Z$$

Explanation

From the definition of *del* above it follows that Z, the output of the gate, is always equal to the nand of the 3 inputs delayed by one time unit.

Comments

Once again it is apparent that a key issue of hardware description is the representation of timing properties. Like HOL, and in contrast to many other languages, temporal logic takes an existing logical framework as its starting point, maintaining a well-defined semantics which will support formal reasoning (see Chapter 5). A mechanised system called Tempura [Moszkowski 85] has been developed to assist in this reasoning process. Whereas higher-order logic has the capacity to describe hardware and its timing properties, temporal logic extends first-order logic with a small set of operators specifically oriented to the description of timing phenomena. It appears that this results in a logic of less complexity than HOL, possibly at the cost of some loss of descriptive power.

2.2.6 Other Languages

It would be an exhausting task to attempt to cover all the current Hardware Description Languages in detail. Even to catalogue them would be almost impossible: a bibliography of HDLs compiled in 1984 [Nash84] contained over 100 references, and the number of languages has dramatically increased since that time. Furthermore, many of the principles of hardware description have already been made clear by the small but diverse selection of languages described in the preceding sections. This Section will therefore provide just a very brief introduction to a few of the more important languages currently in use.

ELLA

Like VHDL, ELLA [Morison 85, Morison 86] has been developed with the aim of serving a large community of users in a wide variety of applications. It also has some similarity to LTS, in that it is based on a strongly-typed, functional programming language. ELLA's model of time is similar to the sequence of discrete instants used in LTS, but a different set of primitive timing constructs is provided. These enable the explicit description of various delay phenomena that are common in hardware, and support the specification of ambiguous delay behaviour. An additional primitive is provided for the description of random access-memory; the aim of this appears to be to increase simulation efficiency. It is noteworthy that the design of the language has been influenced by the implementation of the simulator that it is intended to drive.

μFP

Like LTS [Babiker 83], μFP [Sheeran 83] is based on a functional programming language (in this case, FP [Backus 78]) and hardware is described by functions over sequences of values. By contrast with LTS, however, the semantics of μFP have been formally defined so that correctness proofs may be carried out. An area in which μFP has been successfully applied is in reasoning about highly regular structures, which constitute an important class of VLSI circuits. In particular, μFP has been used to support correctness-preserving transformations, the subject of Section 4.2. In subsequent work, the RUBY language [Sheeran 86, Sheeran 88], in which binary relations between inputs and outputs have replaced functions, has been developed.

CSP

Communicating Sequential Processes (CSP)[Hoare 78] is, like CIRCAL and CCS [Milner 83], a formal, mathematical framework designed to support reasoning about concurrent systems, of which hardware may be considered a subset. This framework has been applied to both the description of hardware and the problem of automated design [Martin 86]. Many of the concepts of the language are similar to those of CIRCAL, which is described in the following Section.

CONLAN

The name of this language [Piloty 82, Piloty 85] comes from 'Consensus Language' which is indicative of the designers' intentions to provide a very general descriptive framework. Rather than defining a single language, they have defined the basis for the construction of a family of new languages from a common starting point, which is Base Con Lan (BCL). Such languages may be set up using CONLAN constructs depending on the application, tool, or level of abstraction for which the language is required. An example of such a derived language is WISLAN [Vaidya 83a, Vaidya 83b] which was used as the input language for a gate-array design system. CONLAN languages have also been used as a front-end for formal verification [Eveking 85b].

Zeus

The language Zeus [German 85, Lieberherr 83, Lieberherr 84] was designed with the goals of encouraging a systematic approach to design and of providing a suitable medium for input to design tools, particularly silicon compilers. It is based on the principles of procedural programming languages, in particular Modula-2 [Wirth 82], of which it is an extension. Behavioural descriptions are written in that language, so there is no convenient way to specify timing information. The language's intended use as input to a compiler is evident from the features that enable behavioural concepts to be associated with pieces of hardware (e.g. states with registers, logical functions with gates, etc.).

In addition to the above languages, several existing frameworks that support formal reasoning have been applied to the description of hardware. HOL, described above, is an example of such an approach. A number of other frameworks will be discussed under the topic of verification in Section 5.2.6.

2.3   CIRCAL

One language, CIRCAL [Milne 83a], will be used in the remainder of this book as the primary example of a hardware description language. There are two main reasons for this choice: firstly, CIRCAL belongs to the subset of languages that support formal reasoning, which is an important topic in this book, and, secondly, it is possible to trace the development of CIRCAL from a very raw descriptive framework to a rich and powerful language. This language was developed by Milne following his work with Milner on concurrent processes [Milne 79] and his subsequent work on the Dot Calculus [Milne 80]. It bears some similarities to Milner's CCS [Milner 80] and SCCS [Milner 83], which also evolved from Milner and Milne's earlier work, and also to CSP [Hoare 78]. All of these frameworks have been developed with the aim of enabling reasoning about the behaviour of concurrent systems, and are based on the idea of processes communicating by a handshaking mechanism. The distinguishing characteristic of CIRCAL among other calculi is that its features have been defined with the specific aim of providing the ability to describe the type of behaviour exhibited by hardware. These features, described in detail below, can be seen to reflect the nature of hardware behaviour as discussed at the start of this Chapter.

2.3.1   *Behavioural Operators*

In CIRCAL the behaviour of devices can be described by assigning expressions to them. This is done using the definition operator '<=', as in

```
GATE1 <=   ( a   CIRCAL expression )
```

GATE1 may be referred to as a *behaviour*.

The *physical* ports of a device are the pins or wires by which it may communicate with other devices. As will be seen below, the ports used in CIRCAL expressions are not always identical to the physical ports of the device. The set of ports to which the expression may refer is called the *sort*. When ports are referred to in the text, they will be identified by an *italic* typeface.

Events and Guarding
One of the key concepts of hardware behaviour is the notion of changes on ports. These are sometimes referred to as *events* (hence the *event-driven* model of VHDL). In CIRCAL events are an essential part of behavioural description. However, although events can be interpreted as changes on ports, this need not always be the case. Some other interpretations are discussed below. The immediately following discussion will make no assumptions about the physical interpretation that might be placed on an event. In general, an event is identified by the port on which it takes place. To differentiate between events and the ports on which they occur, events will be written in a `typewriter` font, and ports in *italics*.

  Multiple events may take place simultaneously. This is represented in CIRCAL by a set of events, which is called a *guard*. For example, if a device has two ports called *ina* and *inb*, and an event takes place on each of these ports at one time, the guard would be represented as

```
{ina, inb}GATE1
```

Behaviours in CIRCAL are most easily interpreted in terms of what will happen if the device's environment (i.e. those components to which it is connected, as pictured in Figure 2.2) tries to perform a certain event or set of events. (The meaning of 'performing an event' will be fully explained below.)

  The interpretation of the above behaviour is therefore as follows: if the environment performs the events `ina` and `inb` simultaneously, they will take place and the subsequent behaviour of the device will be that of `GATE1`. The behaviour of `GATE1` will be defined by some other expression using the operators of CIRCAL.

Deterministic Choice
CIRCAL has two choice operators. The more commonly used one is the deterministic choice operator '+'. It describes alternative behaviours of a device which may be chosen by events occurring in its environment. For example, the behaviour described by the following expression

```
{ina, inb}GATE1 + {inc}GATE0
```

would be interpreted as follows: if the environment performs the two events `ina` and `inb`, then the subsequent behaviour of the device will be that of `GATE1`; if the environment performs the event `inc` then the

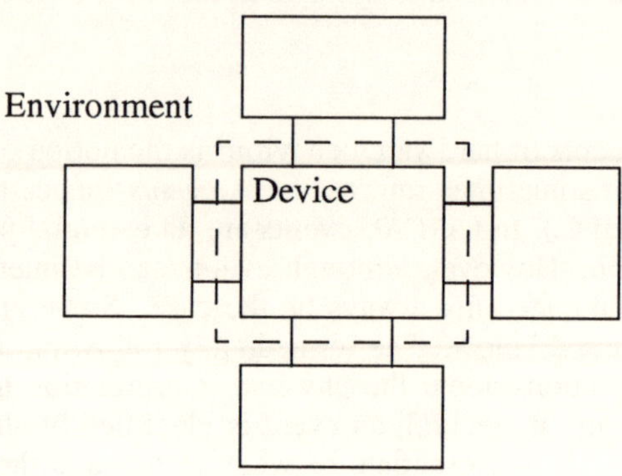

Figure 2.2: A device and its environment

subsequent behaviour of the device will be that of GATE0. The operands
of the choice operator are sometimes referred to as *branches*, reflecting
the tree-like organisation of behavioural descriptions.

For deterministic choice sums involving many terms, the following
notation is sometimes used:

$$\sum_i a_i A_i \;\equiv\; a_0 A_0 + a_1 A_1 + a_2 A_2 + \dots$$

where the a_i's are guards and the A_i's are behavioural expressions.

Nondeterministic Choice

This operator is not often required for the writing of specifications, but
may be necessary to describe the behaviour of a constructed device. It
is represented in some other literature by $\oplus$ but, in order to facilitate its
use in ordinary text, will be represented here by @. Like its deterministic counterpart, it provides alternatives in a behaviour, but the difference
is that the choice between alternatives is made by some unseen action
within the device being described, and is therefore beyond the control
of the environment. Thus, for example, if a device is described by the
following expression:

```
B1 @ B2
```

where B1 and B2 are behaviours, then this means that the device can behave either as B1 or B2, and the environment has no way of determining
which one beforehand. Thus in the following behaviour

```
{ina, inb}GATE1 @ {inc}GATE0
```

an attempt by the environment to perform the event `inc` may or may not succeed; the 'decision' is made within the device, unseen to the environment.

Deadlock
One special behaviour may be required for behavioural description. This behaviour, called the deadlock operator, is represented by Δ, or $\bigwedge$ in ordinary text. It has a sort associated with it, and its behaviour is such that no events can take place on any port in its sort.

The notation introduced above is summarised in Appendix B.

2.3.2   *Describing Hardware Behaviour in* CIRCAL

With the operators presented above it is possible to describe the behaviour of some common pieces of hardware. First of all it is necessary to assign a physical meaning to events. One possible meaning has been mentioned already, a change of value on a port. Suppose there is a physical port called *j* on a device, and that the port can take the values 1 and 0. This could be modelled as a *pair* of imaginary ports, called *j1* and *j0*. Then the event `j1` would be interpreted as 'the value on the port *j* changes to 1.' The following example shows how this approach can be used to describe a delayless inverter.

Example

```
INV0  <=  {in0,out1}INV1
INV1  <=  {in1,out0}INV0
```

Explanation
The two lines of this description can be thought of as representing two states of the inverter. The first line describes the state in which the output holds the value 0. From this state, the device can move into state `INV1` (sometimes referred to as an *end state*) by performing (simultaneously) the events `in0` and `out1`. The second line is the symmetric description for state `INV1`. Note that there is no causality or directionality implied by this description. It simply states that an `in0` event can only happen if accompanied by an `out1` event. This ensures that the output of the inverter, the physical port *out*, always holds the value which is the complement of that on the input, *in*. The ability to describe simultaneous events seen in this example is a distinctive feature of CIRCAL, contrasting for example with CCS [Milner 80].

This type of description is closely related to that used for finite state automata. The relationship between CIRCAL and finite state machine notation is described in [Davie 86b]. Figure 2.3 illustrates the finite state machine representation of the above inverter.

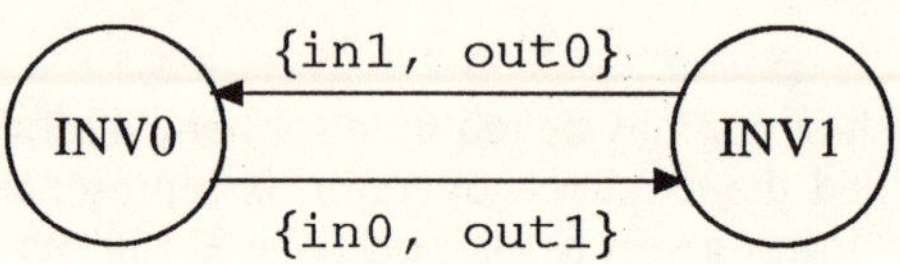

Figure 2.3: Finite state diagram of inverter

Delays
It is easy in CIRCAL to write descriptions in which output events follow input events by some amount of time, as the following example illustrates.

Example
The above description of an inverter can readily be modified to include an unspecified delay between input and output events. One way to do this is as follows:

```
INV01  <=  {in1}INV11
INV11  <=  {out0}INV10 + {in0}INV01
           + {in0, out0}INV00
INV10  <=  {in0}INV00
INV00  <=  {out1}INV01 + {in1}INV10
           + {in1, out1}INV11
```

Explanation
The two digits at the end of each state name represent the values on the inverter's input and output ports. INV01 and INV10 are therefore the stable states; no output change is pending, so the only possibility is that a new value will be presented on the port *in*. In the other two states, there are three possibilities: the output will change, leading to a new stable state; the input change will be reversed, leading back to the original stable state; or both changes will happen at once, leaving the device in the other unstable state. Note that this description makes no statement about how long the delay between input and output events is, only that the complemented input value will appear on the output some time after the input event. Such descriptions are of limited usefulness, as will be demonstrated in Section 5.1.4.

Measuring Time

In order to introduce delays of specified length it is necessary to have some way of measuring time. The standard way of doing this is to introduce a universal clock by which the passage of time can be measured [Traub 86]. This is an abstract device (i.e. it would not be a part of the physical circuit) which simply generates a continuous stream of ticks. These are represented by the event {t}. The time between the ticks is assumed to be constant and would correspond to some physical length of time, e.g. 5 nanoseconds. All timing properties are then measured in terms of this basic unit of time. Note that this approach bears some relationship to that taken in LTS, in which the concept of a sequence of 'instants' was central to the description of timing phenomena.

The convention normally adopted when writing timed CIRCAL descriptions is to assume that events can only take place simultaneously with a 'tick' event. Even though this may not be the case physically, it is a reasonable assumption for modelling purposes providing the time between ticks is sufficiently small. This approach is illustrated in the following description of an inverter with a unit delay (i.e. a delay corresponding to the time between two successive ticks).

Example

```
INV01 <= {in1, t}INV11 + {t}INV01
INV11 <= {out0, t}INV10 + {in0, out0, t}INV00
INV10 <= {in0, t}INV00 + {t}INV10
INV00 <= {out1, t}INV01 + {in1, out1, t}INV11
```

Explanation

As before, the stable states are denoted by `INV01` and `INV10`. In these states, the second term in the choice sum indicates the possibility of a tick happening without any accompanying physical event. In the unstable states, the pending output event is effectively forced to happen on the next tick, either before or at the same time as the next input event. This contrasts with the untimed description above in which there is no guarantee that the pending output event will ever occur.

The above examples serve to demonstrate briefly the principles of behavioural description with CIRCAL for very simple devices. The techniques used and problems encountered in describing more complicated devices are the subject of much of Chapter 3. Before that, however, the structural operators of the language will be described.

2.3.3 Structural Operators

Composition

This is the most important of the structural operators in CIRCAL, being central to the use of the calculus to reason about the behaviour of constructed devices. The composition of two devices corresponds to their being wired together, with the convention that ports of the same name are connected together. Thus, for example, the two devices depicted in Figure 2.4 would be described simply as

```
DEVA  *  DEVB
```

The symbol $*$ is the composition operator, often written as $\bullet$. The sort of DEVA is $\{a, b, c\}$, the sort of DEVB is $\{b, c, d\}$, so that the composition operator joins together ports b and c leaving a and d unconnected. The sort of the constructed device is simply the union of the two sorts, $\{a, b, c, d\}$.

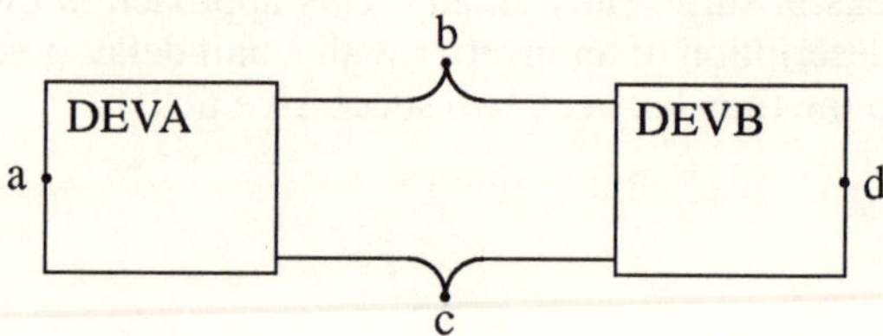

Figure 2.4: Two devices wired together

The 'wiring up' operation can readily be extended to more than two devices. In contrast to CCS [Milner 80] there is no restriction on the number of devices which may be connected to one port, so the composition of several devices connects them together in just the way that one would intuitively expect.

One of the significant features of CIRCAL is that, like higher-order logic, the structural operators have a behavioural meaning. This allows the behaviour of a constructed device to be established mathematically, rather than, say, by simulation. The formal definition of the composition operator is as follows:

Given two devices A and B, with sorts L and N respectively, with behaviours defined as a deterministic choice of guarded terms

$$A \Leftarrow \sum a_i A_i$$

$$B \Leftarrow \sum b_j B_j$$

where a_i and b_j are guards and A_i and B_j are behavioural expressions, then the behaviour of the device obtained by wiring them together is defined by the composition of the two behaviours as follows:

$$A*B \quad \Leftarrow \quad \sum_{a_i \cap N = \phi} a_i (A_i * B)$$

$$+ \sum_{b_j \cap L = \phi} b_j (A * B_j)$$

$$+ \sum_{(b_j \cap L) = (a_i \cap N) \neq \phi} (a_i \cup b_j)(A_i * B_j)$$

$$+ \sum_{(b_j \cap L) = (a_i \cap N) = \phi} (a_i \cup b_j)(A_i * B_j)$$

The four lines of this definition can be readily explained in an intuitive way. The first line represents those parts of the resultant behaviour that arise from guards involving ports which are only in the sort of A. If such a set of events takes place, then A will evolve into the state A_i, but B remains in its original state as its ports were not involved. The second line is the symmetric case for events involving ports that are only in the sort of B.

The third line is really of most interest, as it describes the interaction between the two devices when events occur on ports that are common to both. These *synchronising* events can take place only if the ports of the guard a_i that are in the sort of B and the ports of b_j that are in the sort of A are exactly the same. The resultant guard then is the union of the two guards, and the new state will be the composition of A_i and B_j. Supposing the sort of A is, as before, {a, b, c} and that of B is {b, c, d}. Then, the following synchronisations are possible:

```
{a, b} with {b}
{a, b} with {b, d}
{b, c} with {b, c, d}
```

while the following are not:

```
{a, b} with {c}
{a, c} with {b, d}
{b, c} with {c}
```

The final case to consider is the simultaneous occurrence of events which involve ports only in the sort of A with events which involve only the sort of B. These are not synchronising events, but independent events which happen to occur simultaneously, and it is these which are covered by the last sum in the above definition. The only possibility for such events which exists in the above example is the simultaneous occurrence of {a} and {d}.

Relabelling

In order to describe structure fully, some further operators are required. One of these is the *relabelling* operator, which simply replaces occurrences of a certain port name with a new name. The syntax for relabelling a port called *old* to *new* for a device called DEV is

```
DEV [new/old]
```

This enables the convention of wiring up ports of the same name to be followed. For example, an inverter may be defined to have ports *in* and *out*, and it may be required to connect two such inverters in series as in Figure 2.5. Without relabelling, the wiring up operation would connect them in parallel, joining together the two inputs and the two outputs. To connect them in series, the output of one inverter and the input of the other are relabelled to a common name, and the desired connection is achieved as follows:

```
INV[mid/out] * INV[mid/in]
```

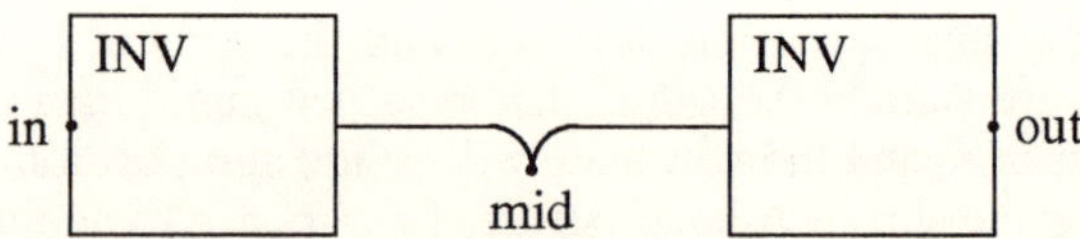

Figure 2.5: Two inverters wired in series

The behavioural meaning of the relabelling operator is quite simple. Any occurrences of a port name designated for relabelling in a behavioural definition are replaced by the new name. In the case where a number of abstract ports are used to represent a single physical port, as was done in the inverter description above, the relabelling can be applied to the physical port name only as a convenient shorthand. For example

```
INV11 [mid/out]
```

would have the following behaviour:

```
INV11 <= {mid0, t}INV10 + {in0, mid0, t}INV00
```

Abstraction

It is sometimes required to internalise some of the ports of a constructed device. This removes some ports from the device's sort, making communication with the outside world through those ports impossible. The analogous situation in VHDL, for example, is to declare certain nodes as signals, rather than externally visible ports, within an architectural body used to describe the structure of a component. To make ports b and c internal to the constructed device of Figure 2.4, the following CIRCAL expression would be used:

```
DEVA * DEVB - b - c
```

The sort of this constructed device would now be `{a,d}`.

In general, the effect of abstracting a port is to remove it from any guards in which it occurs in the behaviour of a device. However, it is possible to introduce non-deterministic behaviour by using abstraction, as the following definition shows:

$$\left[\sum a_i A_i \right] - p \equiv_{def}$$

$$\left[\sum_{p \in a_i \,\wedge\, \{p\} \neq a_i} (a_i \backslash \{p\})(A_i - p) + \sum_{p \notin a_i} a_i (A_i - p) + M \right] \oplus M$$

where

$$M \Leftarrow \sum_{a_i = \{p\}} (A_i - p)$$

The symbol $\sum$ is a non-deterministic choice sum. The subtleties of the non-determinism introduced here are discussed in detail by Traub [Traub 86]. It is sufficient to notice here that non-determinism arises when there are guards consisting of only a single event, on the port which is to be abstracted. These events may now take place unseen and uncontrolled by the device's environment, thus introducing non-determinism into its behaviour. In general, non-deterministic behaviour is something to be avoided. It should be noted that abstraction will not introduce non-determinism if it is not carried out on a port which appears by itself in a guard. In timed descriptions, described below, the only port in this category is the 'tick' port (conventionally t), and it is not usually necessary or appropriate to perform abstraction on that port. Under these circumstances, the definition of abstraction is simplified:

$$\left[\sum a_i A_i \right] - p =$$

$$\sum_{p \in a_i \wedge \{p\} \neq a_i} (a_i \backslash \{p\})(A_i - p) + \sum_{p \notin a_i} a_i (A_i - p)$$

The three structural operators described above are really all that is needed for full structural description of any device. With composition and relabelling, it is possible to describe any desired wiring configuration for a set of devices. Abstraction provides a means to 'black-box' a constructed device, by concealing internal details. All these operators have a behavioural meaning, which provides the means for establishing the behaviour of a constructed device in terms of the behaviours of its components. This is an essential first step in the formal verification process.

2.3.4 *Limits of Pure* CIRCAL

The operators described above form the core of CIRCAL. The language formed from these operators, which will be referred to as pure CIRCAL, can be seen to have severe limitations as a medium for the description of real hardware. A very slightly more complicated example than those seen so far serves to illustrate this.

Example

The following description is of a delayless, 3-input NAND gate, with inputs a, b and c, and an output z, modelled on the assumption that port values may be either 0 or 1:

```
nand000 <= {a1}nand100 + {b1}nand010
        + {c1}nand001 + {a1, b1}nand110
        + {a1, c1}nand101 + {c1, b1}nand011
        + {a1, b1, c1, z0}nand111
nand100 <= {a0}nand000 + {b1}nand110
        + {c1}nand101 + {a0, b1}nand010
        + {a0, c1}nand001 + {c1, b1, z0}nand111
        + {a0, b1, c1}nand011
nand010 <= {a1}nand110 + {b0}nand000
        + {c1}nand011 + {a1, b0}nand100
        + {a1, c1, z0}nand111 + {c1, b0}nand001
        + {a1, b0, c1}nand101
nand110 <= {a0}nand010 + {b0}nand100
        + {c1, z0}nand111 + {a0, b0}nand000
        + {a0, c1}nand011 + {c1, b0}nand101
        + {a0, b0, c1}nand001
nand001 <= {a1}nand101 + {b1}nand011
        + {c0}nand000 + {a1, b1, z0}nand111
```

```
          + {a1, c0}nand100 + {c0, b1}nand010
          + {a1, b1, c0}nand110
nand101 <= {a0}nand001 + {b1,z1}nand111
        + {c0}nand100 + {a0, b1}nand011
        + {a0, c0}nand000 + {c0, b1}nand110
        + {a0, b1, c0}nand010
nand011 <= {a1, z0}nand111 + {b0}nand001
        + {c0}nand010 + {a1, b0}nand101
        + {a1, c0}nand110 + {c0, b0}nand000
        + {a1, b0, c0}nand100
nand111 <= {a0, z1}nand011 + {b0, z1}nand101
        + {c0, z1}nand110 + {a0, b0, z1}nand001
        + {a0, c0, z1}nand010 + {c0,b0,z1}nand100
        + {a0, b0, c0, z1}nand000
```

The fundamental problem of pure CIRCAL that this example illustrates is the lack of any mechanism for handling variables. This means that the states of the device must be explicitly enumerated, leading to a lengthy description in this simple example, and to totally unworkable descriptions for more complex examples. There is no capacity to simplify the description by making use of its regular properties, such as the symmetry between the three inputs. Nor is there any facility to make use of the fact that this device has a very similar behaviour (in terms of the occurrence of input and output events) to many other three-input devices, distinguished only by the particular combinational logic function which it realises.

CIRCAL in its most basic form, then, is not of great practical use as a design and description language. For the description of simple devices, it is rather cumbersome; for devices of reasonable complexity the enumeration of states becomes totally impractical. In order to avoid these problems, the basic language needs to be modified or enhanced in some rigorous way to cope with variables while maintaining the attractive features of CIRCAL such as its support for formal reasoning. Some approaches to this enhancement are discussed in the next Chapter.

These shortcomings of pure CIRCAL, which are certainly surmountable, are not considered a good reason to avoid using it. One of its appealing qualities is that it provides just those features that were identified as being essential for a behavioural description language at the start of this Chapter. This is in stark contrast to many other languages which, in trying to satisfy a very broad range of criteria (e.g. those listed in Section 2.2.1), provide an unwieldy collection of constructs. The provision of a bare minimum of descriptive features in CIRCAL makes it a useful starting point for the development of a language to support formal specification. Furthermore, its simple semantics enable formal reasoning about hardware behaviour, a quality that makes it of particular interest in the validation phase.

2.4   Summary

This Chapter has begun to tackle the issue of hardware description. The formal description of behaviour and structure, at any level in the design hierarchy, is essential to the proposed methodology, regardless of the particular validation method which is to be used. It is this formal description which enables the comparison of specifications with implementations, whether by formal proof or simulation, to take place throughout the design process rather than afterwards. The most viable approach to the description of behaviour is with Hardware Description Languages (HDLs), which have been the subject of much recent attention. These languages are invariably also used for the description of structure, although graphical input of structural information may be desirable from a hardware designer's point of view.

The meaning of hardware behaviour was examined and the concepts identified here led to a set of required criteria for behavioural HDLs. Given the need to describe a variety of hardware over a wide range of levels of abstraction, the following features are required: the ability to describe the interdependency of values on ports; ways to describe various timing phenomena; and facilities to refer to changes occurring on ports.

A small but widely varied selection of languages that met these criteria was examined. VHDL is a language that has been designed to fulfill a great many roles and this is reflected in the extensive range of features that have been incorporated in it during its lengthy development. The features of a procedural programming language enable the description of functions, and timing phenomena are described using a single construct which specifies the length of time between a change of an input and the resultant output change. Type definition capabilities provide some assistance in dealing with complexity at high levels of abstraction.

By contrast, LTS offers the features of a functional programming language much like ML and uses a backward-looking, discrete model of time. The values on ports are represented by signals, which map instants of time to values, and the behaviour of hardware devices is modelled by functions that map signals to signals. The one primitive timing function looks back to the previous instant of discrete time; more elaborate functions can be defined in terms of that one primitive.

The M language, which is intended as a modelling language for functional simulation, was examined. Language extensibility is provided by giving the user full access to the C programming language for type and function definition.

None of these languages supports formal reasoning about hardware. Higher-order logic is a pre-existing framework which does support such reasoning and has been applied to the description of hardware by various people with considerable success. The use of predicates to represent hardware behaviour with a similar timing model to that of LTS was examined and this approach will be addressed again in the chapter on verification.

The importance of being able to describe timing aspects of hardware behaviour was illustrated by temporal logic, which extends conventional logic with a small set of operators that refer to temporal properties. The semantics of this language are defined in such as way as to permit formal verification.

Finally CIRCAL, the language which will used for most examples in this book, was introduced in some detail. This introduction was to the language in its most basic form, which just manages to meet the criteria established for behavioural languages. The well-defined semantics of CIRCAL enable it to be used for formal reasoning about the behaviour of constructed devices. However, its shortcomings with regard to the description of non-trivial devices were illustrated; ways in which these may be overcome will be presented in the following Chapter.

THREE

SPECIFICATION

In Chapter 1 the case was made for an integrated, hierarchical approach to design and validation, which would enable the correctness of each design step to be established before the whole design is completed. It was shown that in order to do this formal specifications must be written at each level of abstraction. The specification task, therefore, is of central importance to this approach; without it, validation cannot be performed until the design reaches the lowest level of abstraction.

In the previous Chapter, some of the issues of behavioural description were addressed. From a discussion of the nature of hardware behaviour, a small set of criteria for behavioural description languages was established. In approaching the task of specification within a hierarchical design and validation methodology, it is generally not sufficient to use a language that meets only those few criteria. For example, since specifications must be written for all devices at every level in the design hierarchy, it is important that the task of writing specifications be as easy as possible, and that the specifications themselves be concise.

It is also important that specifications be accurate, in two senses of the word. The first of these is that they must accurately reflect the real intentions of the designer. The attainment of this goal can never be guaranteed, as it depends on being able to read the mind of the designer. However, it can be assisted by language features that allow 'intuitive' descriptions and save the designer from having to specify those details that do not currently concern him.

The second sense in which specifications must be accurate is that they should, as far as possible, be an accurate description of real hardware. This goal can be met only to a limited extent, since it basically amounts to a requirement that all specifications be implementable. However, a language should certainly help rather than hinder the attainment of this goal.

In general, the languages examined in the preceding Chapter met these added requirements for specification languages to some extent. The exception was CIRCAL, which, in its basic form, offered no high-level aids to conciseness or intuitiveness such as function or datatype definition, and which yields cumbersome descriptions for even simple devices. In order to illustrate some of the issues involved in developing a workable specification language, the following Section contains a discussion of ways to enhance or build on CIRCAL to produce such a language.

The writing of concise, accurate specifications is not just a matter of using an appropriate language. There must also be certain specification techniques that will assist the designer in writing specifications more easily and accurately. The details of these techniques will generally depend on the language in use. The central Section of this Chapter describes some techniques that have been developed for use with the enhanced CIRCAL language.

The Chapter concludes with some proposals for a language which has the dual aims of further reducing the cumbersomeness of writing specifications and of ensuring that the specifications are accurate (in the second of the two senses mentioned above). The second aim is achieved in the following way: descriptions in this 'high-level' language can be translated into the enhanced version of CIRCAL, while the features of the higher-level language serve to restrict the designer's access to CIRCAL in such a way that some of the specification techniques mentioned above are automatically enforced. It is intended that such a language could be used as a 'front-end' to a CIRCAL-based design system.

It should be emphasised that the specification of a device need not always precede its design. A standard cell in a library, for example, might have quite complicated timing characteristics; if such a part is to be used in a design, it must be possible to describe fully those characteristics *after* it is designed. Thus, some of the behavioural characteristics that are specified in this Chapter may represent features of designed parts rather than requirements of parts that are to be designed. These considerations have influenced the development of the specification languages and techniques.

3.1 *Enhanced* CIRCAL

Several researchers have proposed extensions to CIRCAL for a variety of purposes [Milne 84, Milne 88a, Milne 88b, Pezze 87, Traub 86]. In the following Sections, a set of enhancements to CIRCAL which aims to reduce the complexity of behavioural descriptions of real devices will be described. The enhancements have largely been developed as a result of attempts to describe significantly complicated pieces of hardware, such as the simple microprocessor [Gordon 81a] that will be discussed in Chapter 6. The features are introduced by examples and their meaning explained in terms of their relationship to the constructs of pure CIRCAL.

3.1.1   Parameterisation of States

The example of a 3-input NAND gate in Section 2.3.4 illustrated the fundamental problem of pure CIRCAL, which is the exponential growth in the size of descriptions as the number of states of a device increases. In order to avoid lengthy state enumeration it is necessary to introduce some form of parameterisation. This will lead to requirements for other features, as described below.

The NAND gate description could be shortened considerably if, instead of writing the behavioural descriptions of `nand000`, `nand100`, etc., it were possible to write just the one parameterised state description, i.e.

```
nand(p,q,r)  <=  ...
```

This notation is just shorthand for the list of states that appears on the left of the NAND gate description of Section 2.3. In order to map such a description back to pure CIRCAL it is necessary to replace the parameters with actual values 0 and 1. It is assumed that the right hand side will carry sufficient information for the type of the parameters to be inferred.

Having adopted this notation to name states on the left-hand side of behavioural expressions, it becomes necessary to have some similar way to name them on the right-hand side. This could be done in a similar way to the invocation of functions with actual parameters in a programming language. These actual parameters may have literal values (0, 1, true, etc.), may simply be named parameters, or may be functions of other named parameters. Examples of the use of each of these types of parameter assignment appear below. The use of parameter assignment can be more clearly understood once the nature of events in this enhanced language is explained. This is done below.

3.1.2   Value Passing

In pure CIRCAL, events were identified only by port names. To represent new values appearing on physical ports, imaginary ports were defined to correspond to the pairing of physical ports with values. If on a certain physical port, say p, several alternative events are possible, this would normally be represented as:

$$\{pv_1\}S_1 + \{pv_2\}S_2 + \{pv_3\}S_3 + \dots$$

where the v_i are values and S_i are state names. A shorthand way of representing this would be

```
{p~x}N(some parameters)
```

where x is a variable ranging over all the v_i. This will work only if all the end states S_i are just different 'invocations' of one parameterised state N. Fortunately, this is often the case. Consider the following example, which describes a device to input an integer, increment it, then output the incremented value at the next tick of the universal clock.

```
INPUT     <=  {in~x,t}OUTPUT(x+1)
OUTPUT(n) <=  {out~n,t}INPUT
```

Note that the two uses of value passing in this example are different. In the first case, the variable x is bound at the occurrence of the event, and is then used to determine the parameter for the end state. In the second case, the variable is already bound in the start state and is simply presented on the port. Thus it is only in the first case that the shorthand for a choice of alternative events is really being used. In the second, the notation is shorthand for a list of state definitions, as in

```
OUTPUT(0)  <=  {out0,t}INPUT
OUTPUT(1)  <=  {out1,t}INPUT
OUTPUT(2)  <=  {out2,t}INPUT
  etc.
```

This fundamental difference between the two types of value passing has important consequences for synchronisation, as will be seen below. In order to differentiate between the first case, in which a parameter is input, and the second, where it is output, different symbols are used. Arrowheads that indicate whether information is being transferred from port to parameter or *vice versa* were suggested by Traub [Traub 86] and are used here. Thus the above example is now

```
INPUT     <=  {in>x,t}OUTPUT(x+1)
OUTPUT(n) <=  {out<n,t}INPUT
```

Note also the use of a function (addition) to assign the parameter of OUTPUT when it appears as an end state. It may also be desirable to represent output values as functions of state parameters, as the following equivalent description shows:

```
INPUT     <=  {in>x,t}ADDONE(x)
ADDONE(n) <=  {out<(n+1),t}INPUT
```

The appearance of a literal value on a port, which in pure CIRCAL was represented by events such as z1, can be considered as a special case of the outputting of a parameter, in which the output 'function' returns a constant value. Such events are therefore represented in the same manner as outputs, i.e. z<1.

In summary, there are three basic categories of event. These are

- pulses. This word will be used to refer to the basic events of pure CIRCAL in which no value is passed. They are retained in the enhanced version mainly for the description of timing phenomena.
- inputs. Written as p>x where p is a port and x is a variable, this notation represents a range of possible events which result in the binding of x to some value from which parameters of the end state can be derived.
- outputs. Written as p<y where y is in general a function of the parameters of the state whose behaviour is being defined. There are two special cases: the function is a constant, i.e. a literal value is passed, or the function is the identity, i.e. one of the state parameters is passed.

The following definitions express the meaning of the value passing notations more formally. For input:

$$p{>}xS(x) =_{def} \sum_{x_i} \{px_i\}S(x_i)$$

where the x_is range over all possible values for x. This clearly depends on the *type* of x; this issue is addressed below.

For outputs, the statement

$$S(y) \;\texttt{<=}\; \{p{<}f(y)\}R(g(y))$$

is defined to mean

$$\forall y_i.S(y_i) \Leftarrow \{p\,f_i\}R(g(y_i))$$

where

$$f_i = f(y_i)$$

and the y_is range over all possible values for y.

3.1.3 Conditionals

It is not difficult to see the need for conditionals among the enhancements to the language. Having condensed a behavioural description by parameterising the states, it may be required to assign quite different behaviours to certain states. Thus there is a need for a form of conditional which assigns behaviours to states if some predicate on the parameters of the state is satisfied. This is illustrated in the following example.

Example

A counter that counts from 0 up to 7 and then resets to 0 can be described
using conditionals as follows:

```
COUNTER(x) <=
      if eqs(x,7) then {clk,out<0}COUNTER(0)
   + if noteq(x,7) then {clk,out<x+1}COUNTER(x+1)
```

Explanation

eqs() is a function that returns true if its arguments are equal, while
noteq() returns true if they are unequal. The interpretation of the description is that the behaviour of COUNTER for any given state parameters is given by the choice of all branches for which the predicate evaluates to true. In this case, only one branch can evaluate to true for a single
value of x , but in general there will be several. If for a certain set of
state parameters the predicate evaluates to false, then the corresponding
branch is removed from the choice sum. Thus, for example, if x is equal
to 5, then the behaviour of the device is simply

```
{clk,out<6}COUNTER(6)
```

Formally, this notation may be defined as follows:

$$\texttt{if p(X) then B(f(X))} =_{def} \sum_{p(X_i)} B(f(X_i))$$

where

    X is a set of state parameters;
    p(X) is a boolean function of the parameters;
    f(X) is a function mapping the parameters to some new set
of parameters;
    B() is a parameterised behaviour.

Since B() is just a behaviour, the following notation is sometimes
used:

```
if p then ( b1 + b2 + b3 )
```

This is equivalent to

```
     if p then   b1
   + if p then   b2
   + if p then   b3
```

where p is some predicate and b1, b2 and b3 are branches.

There is an additional requirement for predicates, when a parameter is input. When the notation for value passing was introduced, it was mentioned that, in an event written as p>x, x is a variable that can range over the full set of values applicable to a variable with the type of x. There are many cases when it is necessary to define a more restricted set. This is done by attaching a predicate to the parameter. Because this is an assertion to determine the range of possible passed values, rather than a conditional to determine valid branches in the behaviour, a new syntax is adopted:

```
portname > variable : predicate
```

The passed value may now be any value of the appropriate type for which the predicate evaluates to true. The interpretation of this notation is defined as a simple extension of value passing for inputs:

$$\texttt{p>x:f(x)S(x)} =_{def} \sum_{f(x_i)} \{px_i\}S(x_i)$$

with the x_is ranging only over those values for which the predicate $f(x_i)$ evaluates to true.

The following example, a two-input AND gate, illustrates how this notation is used.

Example
```
AND(p,q)  <=
        {a>x:eqs(and(x,q),and(p,q))}AND(x,q)
    +   {b>x:eqs(and(p,x),and(p,q))}AND(p,x)
    +   {a>x,b>y:eqs(and(x,y),and(p,q))}AND(x,y)
    +   {a>x:noteq(and(x,q),and(p,q)),
                        z<and(x,q)}AND(x,q)
    +   {b>x:noteq(and(p,x),and(p,q)),
                        z<and(p,x)}AND(p,x)
    +   {a>x,b>y:noteq(and(x,y),and(p,q)),
                        z<and(x,y)}AND(x,y)
```

Explanation
In each of the first three branches, the predicates on input parameters ensure that any value that is input will be such that the value on the output does not need to change. The last three branches contain predicates which ensure that any new input value will be such that a new output value is required, and so the guard also contains an event on the output

port z. The reason for writing a description in this way is to prevent the generation of 'fictitious' events, a problem which is described in more detail in Section 3.2.1.

The reader may consider that, in terms of complexity, this description is not a great step forward from the pure CIRCAL description of a NAND gate in Section 2.3. This is true, but is a consequence of the fact that this is a simple device with only boolean ports. The reductions of length of description are much more noticeable for more complex devices. Some of the more sophisticated devices described below would be virtually impossible to describe without the notation just presented. In particular, the large example of Chapter 6 illustrates the savings in effort that enhanced CIRCAL offers for the description of complicated circuits.

3.1.4 *Functions and Types*

In the foregoing discussion, three situations requiring the use of arbitrary functions have been seen. These are:

1. to assign parameters in end states;
2. to describe output values in terms of state and input parameters;
3. to define predicates for conditionals.

Some framework for the definition of such functions is clearly required.

It is also necessary to have some means of defining types. The examples chosen so far have only made use of boolean and integer ports and state parameters, but more elaborate types are essential for the definition of more complex devices. As a simple example, for the counter described in Section 3.1.3, it might be helpful to define a type to represent the subrange of integers from 0 to 7, and to define an increment operation on such integers. Or, to describe a memory, it is necessary to have a parameter to represent the state of the memory. One way of doing this would be to define a type mem which would support read and write operations and use a variable of this type for the state parameter.

So, in addition to the features for hardware description developed in this Section, there is a clear need for the common programming language features of function and type definition. Where such features are required in the following examples, the language Standard ML [Harper 86] will be used, since it provides the necessary descriptive power, has well-defined semantics, and is sufficiently intuitive to be fairly understandable even to those not familiar with it.

The features described above will be used for examples throughout the remainder of this book. A formal specification of the syntax of the enhanced language appears in Appendix C. In the following Section, the use of this language in a more complex example is demonstrated.

3.1.5 An Example

The following example is of an 8 bit counter which counts either up or down depending on the polarity of an input called *up*. The device is positive edge-triggered, i.e. it counts when a rising edge occurs on the clock line (the port *clk*). The first step is to write ML function definitions for use in the CIRCAL behavioural description:

```
fun noteq(x,y) = if x = y then false else true;
fun not(x) = if x then false else true;
fun and(a,b) = if a then b else false;
fun incr(x) = if x = 255 then 0 else x+1;
fun decr(x) = if x = 0 then 255 else x-1;
```

The first function returns true if its arguments are not equal. The next two are standard boolean functions, and the last two are increment and decrement functions for integers in the range 0 to 255 (i.e. those which can be represented by 8 bits). The CIRCAL description is now:

```
UDCOUNT(x,c,u) <= if and(u,not(c)) then
  {clk<true, out<incr(x)}UDCOUNT(incr(x),true,u)
+ if and(not(u),not(c)) then
  {clk<true, out<decr(x)}UDCOUNT(decr(x),true,u)
+ if c then {clk<false}UDCOUNT(x,false,u)
+ {up>z:noteq(z,u)}UDCOUNT(x,c,z)
```

Explanation

The state UDCOUNT has parameters x, c and u representing the values on the ports *out*, *clk* and *up* respectively. Thus the first branch states that if the value on the clock port is false, and the value on *up* is true, then a rising edge on the clock will cause the value on the output to be incremented. If u is false, then a rising edge on *clk* will cause the output to be decremented. The last two lines describe events that produce no output change — a falling edge on the clock, and the changing of the value on *up*.

This example exhibits most of the features introduced in the preceding paragraphs and demonstrates the increased power of the enhanced language over pure CIRCAL. UDCOUNT is a parameterised state. On the right hand side of the expression, its parameters are assigned variously to other parameters, functions of parameters, or literal values. Various types of events appear in the example: passing of literal values, the output of functions of state parameters, and the input of parameters. Also, both types of conditional are used. In the last line of the description the predicate noteq ensures that the value appearing on the port *up* is different from the one there initially. The if...then clauses in the other three branches test whether an event on the *clk* port may involve a change to true or false, and in the former case, determine then whether the counter should count up or down.

3.1.6 Composition

One of the attractive features of pure CIRCAL is that it enables the construction of behaviours by use of the composition operator. It is to be hoped that the introduction of additional descriptive power would not remove this capability. Fortunately, this turns out to be the case, although composition in this richer language is certainly more complicated.

In Section 2.3.3 the synchronisation of guards was discussed. The rules for this synchronisation depended on the assumption that events were represented only by port names, and that any two events on the same port could synchronise. In the enhanced language, however, two events on the same port will synchronise only in certain instances. For example, if a port p on one device is able to pass only the value 'true', and the port p on another device is able to input a variable x, then the effect when the two devices are wired together will be to pass the value 'true', this value being input to the second device. However, an event such as `p<true` could not synchronise with the event `p<false` even though they involve the same port, just as in pure CIRCAL the events `in0` and `in1` could not synchronise. A systematic way of determining whether two events can synchronise is required.

There are actually four factors to consider when a pair of events synchronise: whether synchronisation can take place at all; what the resultant event will be; what binding of variables takes place; and what conditional should precede the event (this last consideration is explained below). Some notation will help in the discussion of these factors; to this end, the following functions are defined:

> `port(x)` : given an event x, returns the name of the port on which it occurs.
>
> `synch(x,y)` : returns true if the events x and y can synchronise, false otherwise.
>
> `result(x,y)` : returns the event which is the result of synchronisation of events x and y.
>
> `pred(x,y)` : returns the predicate representing the leading conditional when x and y synchronise.

Four types of events need to be considered: pulses (the events of pure CIRCAL), inputs, outputs (of either functions or parameters), and passing of literal values. The eight possible distinct interactions that can occur are described below. The first named event will be represented by x, the second by `y`. In all cases synchronisation occurs only if `port(x)` = `port(y)`.

1. Pulse – pulse. This is the simple synchronisation of pure CIRCAL. `synch(x,y)` is always true. No variables are bound, no conditional need precede the resultant event.
2. Pulse – anything else. No synchronisation can occur.
3. Input – input. Synchronisation can occur if the predicates associated with the two input parameters are not mutually exclusive. The resultant event is an input, with a predicate on the parameter which is the conjunction of the two original predicates. The two input parameters should be bound to a single parameter. Note that this binding should be carried out not just for the input event, but for all occurrences of that parameter in other events in the guard (e.g the parameter z in `{in>z, out<f1(z)}`) and in the new state that follows the guard. No preceding conditional is required.
4. Input – output. Synchronisation can always occur. The resultant event is identical to the output event. The input parameter should be bound to the output parameter or function. A preceding conditional should be added to ensure that the value that is output satisfies the predicate on the input parameter. For example, if the events that are to synchronise are `p>z:g(z)` and `p<f(a,b)` then `pred(x,y)` = `g(f(a,b))`. The role of `pred(x,y)` in the synchronisation of guards is described below.
5. Input – value. Synchronisation can only occur if the value satisfies the condition on the input parameter. The resultant event is the passing of the value, and this literal value should replace any occurrences of the input parameter. There is no leading conditional.
6. Output – output. Synchronisation can always occur, and the resultant event can be either of the two synchronising events. The leading conditional must be a test for equality between the two passed parameters or functions.
7. Output – value. Synchronisation can always occur, resulting in the passing of a value. The leading conditional is the test for equality between the output parameter or function and the passed value. No variables need be re-bound.
8. Value – value. Synchronisation can only occur if the values are equal, this being equivalent to the synchronisation of pulses. There is no need for a leading conditional or the re-binding of variables.

In general, synchronisation involves not just events but guards. The rules for the synchronisation of guards given in Section 2.3.3 can now be generalised to cover this richer language. Consider two guards a and b defined as

$$a \;=\; \bigcup_i \{x_i\}$$

$$b \;=\; \bigcup_j \{y_j\}$$

where the x_is and y_js represent single events. It is assumed that a is associated with a behaviour whose sort is M and b with one whose sort is N. The symbol $\bigcup$ generates the union of its arguments. Let pa and pb be the set of ports that appear in the guards a and b respectively, i.e.

$$pa \;=\; \bigcup_i \{\mathtt{port}(x_i)\}$$

$$pb \;=\; \bigcup_j \{\mathtt{port}(y_j)\}$$

Then synchronisation of the guards can occur if the following condition holds:

$$pa \cap N = pb \cap M = pc \neq \phi$$
$$\wedge \quad \forall i,j.\,\mathtt{port}(x_i) = \mathtt{port}(y_j) \Rightarrow \mathtt{synch}(x_i,y_j)$$

The guard itself is

$$(\bigcup_{\mathtt{port}(x_i) \notin pc} \{x_i\}) \cup (\bigcup_{\mathtt{port}(y_j) \notin pc} \{y_j\}) \cup (\bigcup_{\mathtt{port}(x_i)=\mathtt{port}(y_j)} \{\mathtt{result}(x_i,y_j)\})$$

The guard must be preceded by a conditional, which is the conjunction of all the predicates generated by the synchronisation of events, as described above. Thus, it is described by the following expression:

$$\bigwedge_{\mathtt{port}(x_i)\,=\,\mathtt{port}(y_j)} \mathtt{pred}(x_i,y_j)$$

where $\bigwedge$ takes the conjunction of its arguments.

Examples
The following examples should serve to clarify the ways in which guards can synchronise in the enhanced language. Let A be a behaviour with sort $\{a,b,c,d\}$ and B a behaviour with sort $\{b,c,d,e\}$. In each of the following examples, the first-mentioned branch forms part of the definition of

A, the second is part of the definition of B. `p1`, `p2`, `p3` are predicates, and `f1` is a function. `eqs` is also a predicate, which evaluates to true if its two arguments are equal. The predicate `and` has the conventional meaning.

- `{b>x:p1(x)}A(x)` can synchronise with `{b<y}B(y)` to give if `p1(y)` then `{b<y} A(y)*B(y)`. (Rule 4)
- `{b>x:p1(x)}A(x)` can synchronise with `{b>y:p2(y)}B(y)` to give `{b>x:and(p1(x),p2(x))}A(x)*B(x)`. (Rule 3)
- `{a<m, b<5, c>x:p3(x)}A(x)` can synchronise with `{b<y, c<7}B(y)` to give if `eqs(y,5)` then `{a<m, b<5, c<7} A(7)*B(5)` only if `p3(7)` evaluates to true. Otherwise no synchronisation can occur. (Rules 7 and 5)
- if `p1(z)` then `{b<z}A(z)` can synchronise with `{b<f1(y)}B(y)` to give if `and(eqs(z,f1(y)),p1(z))` then `b<f1(y)A(y)*B(y)`. (Rule 6)
- `{b>x:p1(x),d<z}A(x)` cannot synchronise with `{b<y}B(y)` simply because *d* is in the sort of B but does not appear in the second guard.
- `{b>x:p1(x),c<3}A(x)` cannot synchronise with `{b<y,c<5}B(y)` because the two events on the port *c* cannot synchronise. (Rule 8)

The fact that synchronisation generally involves guards rather than just single events can sometimes cause problems. Consider the synchronisation between the following two guards:

```
{in>x, out<not(x)} and {out<y}
```

Following the rules for synchronisation presented above, the resultant guard, given that *in* is not in the sort of the second device, should be

```
if eqs(not(x),y) then {in>x, out<not(x)}
```

The trouble with this is that the leading conditional, whose meaning is defined only when it depends on state parameters, now depends on an input parameter as well. In order to turn this into a meaningful expression, the leading conditional should be attached instead to the input parameter to which it refers, as follows:

```
{in>x:eqs(not(x),y), out<not(x)}
```

The general rule for situations like this is that if a leading conditional is generated which refers to an input parameter, it should be moved to qualify the appropriate input parameter. There may be a choice of parameters to qualify, in which case the predicate should be attached to the one that appears last in the guard.

As a final example, consider the synchronisation between the following two guards:

```
{in>x, out<f1(x)} and {in>y, out<f2(y)}
```

Following the event synchronisation rules 3 and 6 gives

```
if eqs(f1(x),f2(x)) then {in>x, out<f1(x)}
```

with the input parameter y being bound to x throughout (including the leading conditional). By the argument of the previous paragraph, this becomes

```
{in>x:eqs(f1(x),f2(x)), out<f1(x)}
```

This concludes the introduction to the enhanced CIRCAL language. The following Section illustrates some of the techniques that assist the designer in using the language for the writing of specifications in a design and validation methodology.

3.2 Specification Techniques

The enhanced CIRCAL language has been introduced to provide more descriptive power and greater ease of description than is provided by pure CIRCAL. The task of writing specifications can be further eased by using certain techniques with the language. Certain techniques may also be required to improve the accuracy of descriptions. A number of such techniques have been developed through experience of using the language and are presented below. The motivation for easing the specification task is not simply laziness: in the proposed approach to design and validation, specifications must be written at every level of hierarchy, and so their writing may constitute a significant portion of the design effort. Furthermore, a reduction in the effort of writing descriptions improves a designer's chances of correctly formalising his informal ideas about a device's behaviour.

Some of the following techniques have arisen from attempts to perform verification, as it is sometimes only when one tries to make practical use of a specification that its shortcomings become apparent. This issue is treated more fully in Chapter 5. Other techniques have been motivated by the need to use the CIRCAL modelling framework in a meaningful way, as is discussed in the next Section.

3.2.1   *Specifying Valid Events*

One of the first issues that arises when writing specifications in enhanced CIRCAL, which has already been seen in some of the foregoing examples, is the need to ensure that events really do represent changes in the values on ports. As was mentioned before, it is not really meaningful to allow an event such as p<3 if the value on the port *p* is already 3. For input events, the way to deal with this is to have a state parameter corresponding to the current value on the port, then attach a predicate to the input parameter to guarantee that it is different from the current value. The following example demonstrates this.

Example

```
A(x)  <=  {p>y:noteq(y,x)}  A(y)
```

Explanation

The function noteq(x,y) was defined above to return true if its arguments differed. This device will therefore only accept events on the port *p* if the value differs from that already there, represented in this case by the state parameter x.

For output events a fairly similar technique can be adopted. A state parameter is again required to retain the current value on the output port, and then some test of various state and input parameters must be applied to decide whether the output will change or not. The exact nature of this test depends on the type of device being described. A simple example will illustrate the technique.

Example

The device described below is intended to transfer input values to the output after a delay of two time units (ticks).

```
DEL(x,y,z) <= if eqs(y,z) then
    ({t, in>p:noteq(p,x)}DEL(p,x,y)
     + {t}DEL(x,x,y))
  + if noteq(y,z) then
    ({t, in>p:noteq(p,x), out<y}DEL(p,x,y)
     + {t, out<y}DEL(x,x,y))
```

Explanation

The parameter z represents the current value on the output port, and y represents the value that will appear there on the next tick. If these are equal, therefore, the next tick need not be accompanied by an output event, as indicated by the first two branches. If y differs from z, as the predicate in the last two branches ensures, then an output event will be required.

The question arises of what would happen if a specification were written which did not ensure that events really did represent changes of value. Certainly if the descriptions were to be simulated there would be a penalty in efficiency, as the number of events occurring could be greatly increased. The consequences for verification are a little more involved and will be discussed in Chapter 5.

3.2.2 Constructive Specification

Because the structural operators of CIRCAL have a well-defined behavioural meaning, it is possible to use them to construct a specification. This may be useful, for example, if a device to be specified has a number of fairly independent functions. It is then possible to write the specification as if the device were constructed from several parts, each one performing one of the functions. The splitting of a specification in this way does not necessarily commit the designer to implementing the device in the same way. An example of a device which logically may be split into two parts appears below.

Example

The following specification describes a counter with an asynchronous load capability. Its two functions, counting and loading, are described separately by the behaviours COUNT and LOAD.

```
LOAD(false,n,m) <=  if noteq(m,n) then
   ( {asl<true, int<n}LOAD(true,n,n)
    + {asl<true, int<n, data>x:noteq(x,n)}
                     LOAD(true,x,n) )
  + if eqs(m,n) then
    ( {asl<true}LOAD(true,n,n)
     + {asl<true,data>x:noteq(x,n)}
                     LOAD(true,x,n) )
  + {data>x:noteq(x,n)}LOAD(false,x,m)
LOAD(true,n,m) <= {asl<false}LOAD(false,n,m)
  + {data>x:noteq(x,n)}LOAD(true,x,m)
  + {asl<false, data>x:noteq(x,n)}
                     LOAD(false,x,m)
COUNT(false,p) <= {clk<true, out<incr(p)}
                     COUNT(true,incr(p))
  + {int>x:noteq(x,p)}COUNT(false,x)
  + {clk<true, out<incr(p),int>x:noteq(x,p)}
                     COUNT(true,x)
```

```
COUNT(true,p)  <= {clk<false}COUNT(false,p)
  + {int>x:noteq(x,p)}COUNT(true,x)
  + {clk<false,int>x:noteq(x,p)}COUNT(false,x)
COUNTER <= LOAD * COUNT - int
```

Explanation

The behaviour LOAD is parameterised over the values on the ports *asl* (asynchronous load), *data*, the port for input of new values, and *int*, the port for communicating these values to COUNT. Loading is triggered by a rising edge on *asl*, causing the value on the *data* port n to be transferred to *int* if it differs from the value there already. The parameters of COUNT are the value on the *clk* port and the stored count value p. A rising edge on *clk* will increment p, or it may be overwritten at any time by a new value being communicated on *int*.

Of course, the counter would probably not be implemented in two parts like this. However, writing the specification in two parts makes the whole task rather simpler because it removes the need for the designer to specify the possible interleaving of events on the *asl* and *data* ports with those on the *clk* port. This simplification improves the chances that the designer's intentions will actually be captured by the specification.

The parts that make up a constructed specification need not even be physically realisable — their sole function is to simplify the writing of a specification. For example, it is common to write specifications in which only rising or falling edges of a signal are important. Such specifications can be somewhat simplified by the use of an edge detector box, as in the following example.

Example

An edge-triggered D latch may be described in the following way:

```
DLAT <= D * EDET   - out
EDET(0) <= {clk<1, out}EDET(1)
EDET(1) <= {clk<0}EDET(0)
D(x,z) <= {data>y:noteq(y,x)}D(y,z)
   + if noteq(x,z) then
    ( {out, q<x}D(x,x)
      + {data>y:noteq(y,x), out, q<x}D(y,x) )
   + if eqs(x,z) then
    ( {out}D(x,z)
      + {data>y:noteq(y,x), out}D(y,z) )
```

Explanation

The edge-detector box, EDET, produces a pulse on *out* only when a rising edge occurs on *clk*. The behaviour of D is parameterised over x and z, the values on the *data* (input) and *q* (output) ports respectively. A new value may always be placed on *data*. If x and z differ, then a rising edge on the clock, denoted by the out event, will cause x to be placed on the output; otherwise it causes no change.

The main advantage of this approach is that it saves the designer having to take account of all possible interleavings of rising and falling clock edges with other possible events. This fairly small saving in effort becomes considerable if many edge-triggered devices are used, since the edge detector need only be specified once, and can then be used in all cases, suitably relabelled. Also, the amount of effort required to specify interleavings of events increases with the number of input ports, so the savings of this approach are more apparent. More generally, a designer could build up a library of useful specifications to be used in constructing the specifications of complex devices. More examples of the types of specifications that such a library might include are described below.

Constructive specification is also used for the addition of constraints to a specification. In this situation, the composition operator can be seen to behave like conjunction in higher-order logic, which is also the 'wiring-up' operator in that framework. This aspect of constructive specification is described in Chapter 7.

3.2.3 Delays

The specification of timing phenomena has been seen to be one of the important aspects of hardware description, and in particular the specification of delays is frequently required. There are two basic classes of delay: transport, where a change on an output follows a change on an input by a certain length of time, and inertial, where a change on an output will only occur after an input has maintained its new value for a certain period of time. It may also be desired to specify delays that are of differing lengths depending on the direction of the input change or the port on which it occurs. CIRCAL has several ways of treating these types of delay, which are discussed below.

The simplest way to describe a delay in CIRCAL is illustrated in the following example, which is an inverter with a delay of two ticks.

Example
```
INV(x)  <= {t,  in>y:noteq(y,x)} {t}
                    {t,  out<not(y)}INV(y)
   + {t}  INV(x)
```

Explanation
The sequence of three guards here means that on the first tick a new input arrives, then another tick happens, then on the third tick a new output is presented. The problem with this description is that there is a period of

time during which no further input events can occur. This is an example of an 'inaccurate' description, in the second sense mentioned at the start of this Chapter. It is contrary to the normal understanding of hardware behaviour, where inputs are assumed to be always able to accept inputs, even if no action is taken until some later time such as the occurrence of a clock. The consequences of this approach to specification can be serious when verification is attempted, as is discussed in Chapter 5.

A better approach, then, would always allow input events, but keep track of the pending output changes. This can be done by using some extra state parameters which act like a queue of values, from which the new output value eventually pops off the end. The description now becomes

```
INV(x,y,z) <= if eqs(y,z) then
    ({t}INV(x,not(x),y)
      + {t, in>p:noteq(x,p)}INV(p,not(x),y))
  + if noteq(y,z) then
    ({t, out<y}INV(x,not(x),y)
      + {t,in>p:noteq(x,p),out<y}INV(p,not(x),y))
```

Explanation

The parameter y represents the pending output value. If it equals z, the value on the output, then a tick will produce no change. However, a tick may be accompanied by an input event on *in*, which replaces the value on that port (represented by x) with a new value, p.

The above description is for transport delay. A little modification would provide inertial delay — this is explained below. For longer delays, extra state parameters are needed to provide a longer queue of pending outputs.

Yet another approach, and probably the best, makes use of the constructive specification technique. As with the edge detector example, the saving of effort becomes considerable when a number of similar but not identical devices are being described. The method here is to describe the gate as a delay-less device and compose it with a separate delay device, as in Figure 3.1(a). Thus the two-tick-delay inverter is now described as follows:

```
INV <= INVERT * DEL2 - z
INVERT(x) <= {in>y:noteq(y,x), z<not(y)}INV(y)
DEL2(p,q,r) <= if eqs(q,r) then ({t}DEL2(p,p,q)
      + {t, z>m:noteq(p,m)}DEL2(m,p,q))
      + if noteq(q,r) then ({t, out<q}DEL2(p,p,q)
      + {t, z>m:noteq(p,m), out<q}DEL2(m,p,q))
```

The behaviour of DEL2 is identical to that of the device DEL described
in Section 3.2.1.

For inertial delay, this approach is equally simple. All that is required
is to change the definition of DEL2 to

```
DEL2(x,2)  <=
   {t}DEL2(x,1) + {t, z>y:noteq(x,y)}DEL2(y,2)
DEL2(x,1)  <=
   {t, out<x}DEL2(x,0)
 + {t, z>y:noteq(x,y)}DEL2(y,2)
DEL2(x,0)  <=
   {t}DEL2(x,0) + {t, z>y:noteq(x,y)}DEL2(y,2)
```

Explanation

This type of delay box starts a count-down (the second state parameter) when a new input arrives, as the second branch of each line shows.
The count is reset if another new input arrives before the new output has
propagated. Otherwise, after 2 ticks, the new value is placed on the output. This is a situation where it is important to ensure that events really
do represent changes in values, as discussed in Section 3.2.1, since the
count should only be reset for a genuine new input value.

Using these delay specification techniques, it is possible to combine
inertial and transport delay, a situation which may sometimes be found
in hardware, simply by cascading the appropriate delay boxes. The ease
with which the delay characteristics of a device specification can be modified and the amount of repetitive description that is saved over a large
design are key advantages of this technique. It also reduces the opportunity for errors, since once a delay box has correctly specified for one
device it can be re-used for any number of others.

In some situations, the above technique is not sufficiently general to
describe a device's delays. If there is some asymmetry in the delay characteristics of a device which depends on the direction of input changes or,
for multi-input devices, on the actual input port that causes the change,
then the behaviour cannot be modelled by the simple attachment of a delay box to the output of an ideal device. What is required here is separate
delay boxes on the input lines, as illustrated in Figure 3.1(b). By defining
a 3 unit delay box as

```
DEL3(n,p,q,r)  <= if eqs(q,r) then
   ({t}DEL3(n,n,p,q)
      + {t, z>m:noteq(p,m)}DEL3(m,n,p,q))
  + if noteq(q,r) then ({t, out1<q}DEL3(n,n,p,q)
      + {t, z>m:noteq(p,m), out1<q}DEL3(m,n,p,q))
```

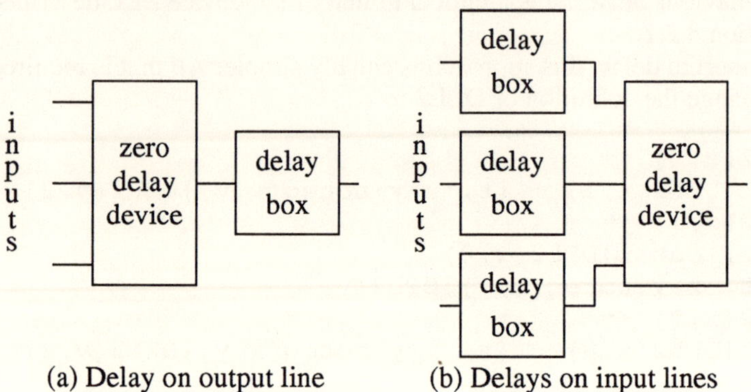

(a) Delay on output line   (b) Delays on input lines

Figure 3.1: Uses of delay boxes

it would be possible to describe a 2-input AND gate which had a delay
of 2 units on signals on the *a* input and a delay of 3 units on the *b* input
as follows:

```
andgate <=
  DEL2 [a/z] * DEL3 [b/z] * AND [out/a][out1/b]
     - out - out1
```

Propagation delays that depend on the direction of change of the input
are quite common. For example, a delay of length 3 units on rising edges
and of length 2 units on falling edges could be modelled by a box such
as the following:

```
DEL2_3(n,p,q,r) <=
    if and(eqs(q,r),eqs(n,false)) then
      {t, z<true)}DEL2_3(true,n,p,q)
  + if and(eqs(q,r),eqs(n,true)) then
      {t, z<false)}DEL2_3(false,false,p,q)
  + if eqs(q,r) then {t}DEL2_3(n,n,p,q)
  + if and(noteq(q,r),eqs(n,false)) then
      {t, z<true, out1<q}DEL2_3(true,n,p,q)
  + if and(noteq(q,r),eqs(n,true)) then
      {t, z<false, out1<q}DEL2_3(false,false,p,q)
  + if noteq(q,r) then
      {t, out1<q}DEL2_3(n,n,p,q)
```

This device behaves very much like a 3 delay box except that when a
falling edge occurs on its input, the value 'false' is placed in the first

2 spots in the queue, rather than just the first spot. This ensures that a falling edge reaches the output in just 2 ticks, while rising edges still take 3 ticks to propagate.

3.2.4   Other Timing Phenomena

Sometimes it is required to describe timing characteristics other than propagation delays. Examples of these are setup and hold times for clocked devices (the length of time before and after the clock edge for which the data inputs must be stable) and the rise and fall times of signals. These types of characteristics are not normally specified exactly, but as ranges of acceptable values. In order to deal with the imprecision of these situations another constructive specification technique can be used. In this case, an imaginary device that generates 'ticks' at the beginning and end of the required time period can be used. The following example illustrates how this works.

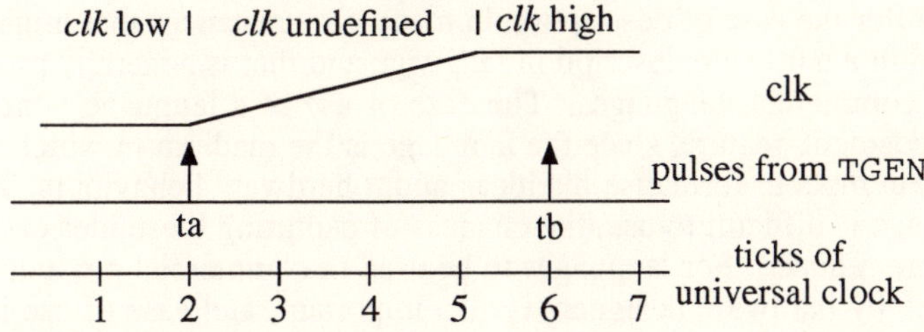

Figure 3.2: Events surrounding rise of clock

Example

Suppose a device has a clock input on a port *clk* and that rise time for signals on this port must be less than 4 ticks of the universal clock. Then a device is defined to generate two pulses, say ta and tb separated by 4 time units:

```
TGEN <= {ta, t}{t}{t}{t}{tb, t}TGEN + {t}TGEN
```

Explanation

The second branch of this behaviour allows TGEN to accept ticks passively; in the first branch, a pulse on *ta* must be followed by 3 ticks of the universal clock then a pulse on *tb* must accompany the 4th such tick.

Assuming that the clock line goes from low, through undefined, to high, then the behaviour of the device can now be described as follows:

```
datatype tristate = low | undef | high

DEV <= DEVA * TGEN - {ta, tb}
DEVA <=
    {clk<low}{ta, clk<undef}{clk<high}{tb}DEVA
```

By writing the specification in this way the clock line is only constrained to go high within 4 ticks after it becomes undefined. The ordering of events that is imposed by this description is shown in Figure 3.2. Similar techniques can be used to describe other timing characteristics that involve a range of timespans rather than a precise length of time. Some of these phenomena relate to the specification of constraints, which is discussed in Chapter 7.

3.3 SuperC

It has probably become apparent by now that, even in its enhanced form and using the techniques developed in the previous Section, CIRCAL does not offer the ease of description in many circumstances that might be desired in a hardware description language and that is generally provided by commercial languages. The ease of use of a language is not a purely cosmetic feature, since the language is the medium in which a designer attempts to formalise his ideas about hardware behaviours. If the language is difficult to use, the chances of capturing these ideas accurately are reduced. For languages to be used in commercial products, acceptance by hardware designers is also important, and ease of use is clearly important here.

The difficulty of using CIRCAL arises mainly because of the event-driven model that underlies the language and which requires the designer to specify explicitly every possible event that can occur in every state of a device. Additional effort is required to test input and output events to ensure that they are 'real' events representing changes in port values.

Another shortcoming of CIRCAL is the ease with which unrealistic devices can be specified. For example, it was shown in Section 3.2.3 that a device can be specified which for certain periods cannot accept input events. Such a device really contradicts the normal understanding of an input. Because an input is not driven by the device to which it is attached, an attempt by another device to change its value cannot be resisted. Even clocked devices, which do not *process* inputs during certain periods, can still accept changes on their input ports during these times. Furthermore, such a specification may lead to erroneous conclusions when validation is attempted.

It would be quite difficult to overcome these problems by adding further features to enhanced CIRCAL. What is required instead is a way to restrict the manner in which CIRCAL is used. This can be achieved by defining a new language that is easily translatable into CIRCAL. The language should aim both to restrict the designer by preventing the specification of unrealistic devices and to reduce the effort of writing specifications. Such a language, which will be called SuperC, is proposed in this Section.

The relationship between pure CIRCAL, enhanced CIRCAL and SuperC is illustrated in Figure 3.3. Whereas enhanced CIRCAL is an extension of pure CIRCAL that can be mapped into it, SuperC is a separate language giving access to only part of enhanced CIRCAL. In general, expressions in enhanced CIRCAL cannot be translated in to SuperC. Therefore, a useful application of SuperC might be as a 'front-end' input language to a CIRCAL validation system; specifications could be more easily written in SuperC, then translated to CIRCAL for subsequent manipulation and comparison.

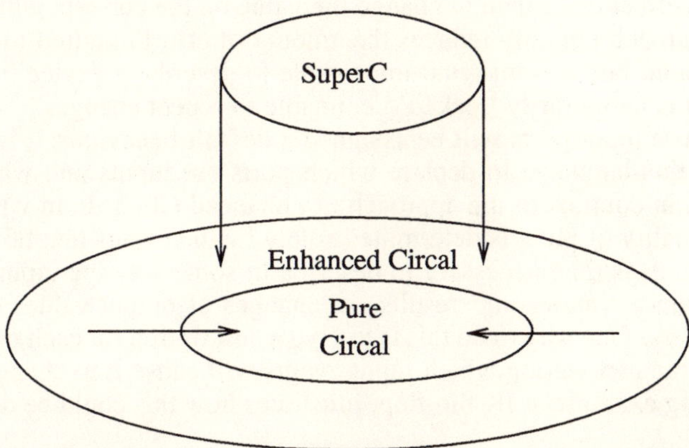

Figure 3.3: Relationship between pure CIRCAL, enhanced CIRCAL and SuperC.

It may be asked whether another language is really necessary; might it not be possible to translate from an existing language into CIRCAL? Certainly it is possible to translate from subsets of some other languages to CIRCAL, but there are at least two advantages in defining a new language: the translation may be effected more easily, and some of the distinctive features of CIRCAL can be retained. In particular, the ability to specify constraints and constrained behaviours is a feature that few other languages offer. This feature is of considerable importance and will be discussed in Chapter 7. In the discussion of the development of the language that follows, some trade-offs have been made between provision of a familiar, programming-language-like syntax and retention of the important features of CIRCAL. In general, the latter was given the higher priority, the intention being to illustrate some of the issues involved in developing a usable design language with well-defined semantics.

3.3.1 Development of the Language

An initial observation that led to the development of this language was that many CIRCAL hardware specifications look very much alike. This similarity is a result of the fact that a large part of any specification is concerned only with the accepting of input events. In many circumstances, the sole effect of an input event is to put a new value on the appropriate port. A number of input events (often a small subset of the total) also cause output events or changes of state parameters. So, the effort of writing descriptions could be reduced by defining the language in such a way that only those input events that produce 'interesting' results need to be specified. All other input events are assumed to be possible and to have no effect other than to change the value on the corresponding port. This approach not only reduces the amount of effort required to write a description, but also makes it impossible to describe a device in which an input is temporarily 'locked' i.e. unable to accept changes.

Because input ports will be assigned a default behaviour, it is necessary in this language to declare which ports are inputs and which are outputs, in contrast to the approach of enhanced CIRCAL, in which the directionality of ports is determined solely by the events that take place on them. It is then necessary to describe in some way the input events that produce 'interesting' results i.e. changes of output values or state parameters. One way to do this is to give a description for each output or state parameter stating which input events will cause it to change. The following example, a JK flip-flop, illustrates how this could be done.

Example

```
jkff(j,k,clk) = q
  where q =
    if and(j,k) then event(clk,true):not(q),
    if and(j,not(k)) then event(clk,true):true,
    if and(not(j),k) then event(clk,true):false,
  end;
```

Explanation

In keeping with many other hardware description languages, the piece of hardware is defined using function notation, where the inputs (j, k and clk) are declared as the arguments and the outputs (in this case just q) as the returned result. In the body of the description, the combinations of port values and input events that will result in output changes are listed. So, for example, the third line above states that, if the values on the j and k ports are both true, then a rising edge on the *clk* port, denoted by `event(clk,true)`, will result in the new value of q being the complement of its old value. Note that the values on ports that are referenced on the right hand side of an equation are assumed to be those which prevailed immediately before the designated event.

Before discussing how such a description could be translated into Circal, it is helpful to consider another example, in which the device being described has internal states, i.e. states not directly visible as values on ports. The behaviour of such states can be represented just as outputs are. The only need for different treatment is in the first line of the description, so that the state is declared in a way that distinguishes it from an input or output.

Example
A random access memory could be described as:

```
memory(mst)(read,write,data,addr) = out
  where
    out = event(read,true):fetch(mst,addr),
    mst =
      event(write,true):store(mst,addr,data)
  end;
```

Explanation
The functions `fetch` and `store` would be defined using a suitable programming language such as ML, just as in enhanced CIRCAL. (These definitions appear in Appendix D.) In the above description, the line beginning 'out =' describes the behaviour of the output port: if the value on *read* becomes true, then the value on *out* becomes the value returned by `fetch`. The following line implies that the internal memory state variable *mst* will be assigned a new value, given by the memory-updating function `store`, if the value on *write* becomes true.

3.3.2 Translation
In this Section, a possible procedure for translating SuperC descriptions into enhanced CIRCAL is described. It is important that such a procedure can be performed so that expressions in SuperC can be used for verification or simulation purposes. As a starting point, it is assumed that a set of data structures has been built up by a parser. A possible set is:

`partname:` A string representing the name of the part (e.g. `"jkff"`).

`inputport_list`: A list of strings corresponding to the inputs (e.g. `["j", "k", "clk"]`).

`outputport_list`: A list of strings corresponding to the outputs (e.g. `["q"]`).

`state_list`: A list of the internal states (e.g. `["mst"]`).

`input_output_list`: A list of the relationships defined between input events and output events. Each element of the list corresponds to one line of the description body and contains three components: an input event, a condition, and an output event. For example, the line `if and(j,k) then event(clk,true):not(q)` from the above flip-flop description would be represented by (`"clk <true"`, `"and(j,k)"`,`"q<not(q)"`). Note that an 'input' event may actually involve the passing of a literal value, as in this example.

`input_state_list`: Identical to the `input_output_list` except that the output events are replaced by state assignments. Assignment could be represented by the character ~, so the penultimate line of the memory description would be represented by (`"write< true"`, `""`, `"mst~store(mst,addr,data)"`), the empty string indicating the absence of a conditional.

Several assumptions may be made about the notation for the resulting CIRCAL description. The name of the device is unchanged, and its state parameters are the union of three sets: the values on the inputs, the values on the outputs, and the declared internal states. The names of the parameters representing the port values are the same as the corresponding port names. The translation procedure could then consist of the following steps:

1. Explicitly add implied information to each element of the `input _output_list` and `input_state_list`. For example, the event `clk<true` is only possible if the value on the port *clk* does not equal true already. Thus the conditional for the first element of the list must be modified to ensure this, becoming `and(noteq( clk,true),and(j,k))`. In general, when the input event is of the form `port<value`, it is necessary to conjoin the existing predicate with a predicate of the form `noteq(port,value)`. If the input events is of the form `port>param`, the event needs to be guaranteed to be valid in the sense defined in Section 3.2.1; this requires the addition of a predicate to the input parameter, so that the input event becomes `port>param:noteq(param,port)`.

2. Ensure that output events are also valid. This involves adding another term to the conditionals in `input_output_list` to ensure that the new output value differs from the value currently on

the port. If the output event is of the form `port<value` then the conditional should be conjoined with a predicate of the form `noteq(port,value)`, thereby ensuring that the output event will actually constitute a change in the value on the port.

3. Having completely processed the list of 'interesting' events in the above way, compile a list of the other events, which cause no output changes and whose only effect on the state is to modify values on input ports. These can be divided into two classes: those that occur on inputs that are not mentioned in the body of the SuperC description (e.g. *j* and *k*), and those that occur on the ports that are mentioned, but cause no output or state change because of the value that is passed or the failure of a conditional to be satisfied. In this latter category for the JK flip-flop are events such as `clk<false`, or `clk<true` when none of the three conditionals is true. Events in the first category are simply of the form `j>j1:noteq(j,j1)`.

4. A complete list of all possible *single* input events and their effects on outputs or internal states has now been compiled from the SuperC description. From this list, construct a list of all possible sets of input events that may occur simultaneously. This is done by constructing the set of all subsets of the original list, excluding those subsets in which different input events on a single port appear in two members of the subset.

5. From each element of this list of subsets, generate a line of enhanced CIRCAL. The input events of all the members of the subset and any associated output events are formed into a guard. The conditionals of all the members of the subset are conjoined to form a single conditional. The effect on the states that represent port values can be easily deduced from the input and output events, while the effect on internal states is as specified in the original `input_state_list` data structure.

The complete CIRCAL description is the deterministic choice of the lines generated in this way.

Some additional features are either essential or desirable for this language. These are presented below.

3.3.3 Combinational Logic

Devices whose outputs depend only on the values of the inputs at a single instant in time, i.e. combinational logic, can be described using the language features presented above. This approach is not as concise as might be desired, however, and the frequency with which such devices are used warrants some special treatment. A special static assignment operator could be used, as illustrated in the following example:

Example
```
andgate(a,b) = c
 where c := and(a,b)
  end;
```

where the function and would be suitably defined in the chosen programming language. In general, an expression of the form

```
z := f(a,b,...)
```

is equivalent to

```
z = event(a,X):f(X,b,...),
     event(b,X):f(a,X,...),
     event(...
         .
         .
     etc.
```

which can be translated to enhanced CIRCAL as before.

3.3.4   *Timing properties*

In Section 3.2.3, considerable attention was given to the problem of specifying delays. Rather than forcing a designer to deal with such concepts as ticks and queues of values, a language should allow him to work directly with the more familiar ideas of inertial and transport propagation delays. Section 3.2.3 showed a number of techniques for the description of these phenomena. These techniques can readily be incorporated into the SuperC language, as the following example illustrates:

Example
In order to specify the above AND gate with a transport delay of 3 time units, the following notation could be used:

```
andgate(a,b) = c
 where c := and(a,b)   tdelay 3
  end;
```

Explanation
The output c takes the value calculated by applying the function and to the values on the ports a and b 3 time units after a change occurs on either of those inputs.

   This description would translate to enhanced CIRCAL as

```
andgate <= and * del3 - z
```

Where `del3` is a delay box as defined in Section 3.2.3, and `and` is the delayless AND gate as before. Inertial delays could be described by using a different reserved word (`idelay`) and the translation would simply involve a suitably defined inertial delay box for `del3`, such as was illustrated in Section 3.2.3.

The description of delays that differ depending on the input port at which the change occurs or the polarity of the change could not easily be incorporated into descriptions using the static assignment (`:=`) operator. It would be necessary to use the full event-based notation presented at the start of this Section and to specify the various delays explicitly. Consider again the description of an AND gate that has a propagation delay of 2 units after changes on the *a* input and a delay of 3 units after changes on the *b* input. This could be described as follows:

```
andgate(a,b) = c
  where c =
    if eqs(b,true) then event(a,x):x  tdelay 2,
    if eqs(a,true) then event(b,x):x  tdelay 3
  end;
```

Explanation

The second line states that the value on the output *c* will change to x, the value input on *a*, if the value on *b* is already true, and that this change will occur 2 time units after the change on *a*. The next line describes the case when *a* is already true and a new value x is input on *b*, the only difference being that the delay is now 3 units.

This is exactly the situation that was described in Section 3.2.3 and would be translated to give the same enhanced CIRCAL specification using a separate delay box for each of the input lines.

3.3.5 *Sequential Behaviour*

An assumption that has been made for all the descriptions so far is that an output event will always be triggered by an input event, possibly lagging behind it by some period of delay. There are, however, many devices for which this assumption is not valid, mainly those that have an internal clock. In such devices, output events are essentially triggered by the passage of time; therefore a way of representing the passage of time is required. A simple if slightly inelegant way to do this is to adopt the CIRCAL approach of a universal clock that generates regularly spaced ticks and to use these as the 'input event' to trigger the change in output. The following example illustrates this approach.

Example

A non-overlapping two-phase clock is described as:

```
twophase(onenext)() = (phione,phitwo)
  where
   phione =
      if and(onenext,not(phione)) then tick:true,
      if phione then tick:false,
   phitwo =
      if and(not(onenext),not(phitwo)) then
                                    tick:true,
      if phitwo then tick:false,
   onenext =
      if phitwo then tick:true,
      if phione then tick:false
  end;
```

Explanation

In the first line, the empty parentheses show that there are no inputs to this device. The outputs are *phione* and *phitwo*. onenext is an internal state that determines which clock phase should rise next. The two lines after 'where' describe the behaviour of the first clock phase: if it is false, and the internal state is true, then a tick should cause it to rise; if it is true, then a tick should cause it to fall. The next two lines describe the second phase: it is identical, except that it rises when the internal state is false. Finally, the internal state's behaviour is defined: it becomes true when the value on *phitwo* is true, and false when the value on *phione* is true. It should be noted that all the output events occur not as a result of input events, as has been the case in other descriptions, but as the result of ticks.

Note that translation of this type of description is virtually identical to the procedure described in Section 3.3.2. The only modification is that the notation tick: translates to a 'pulse' event, denoted by {tick} in enhanced CIRCAL.

3.3.6 *Constrained Specifications*

Another assumption that has been made throughout the discussion of this language is that all devices will be able to accept any input event at any time. Bearing in mind the fact that specifications must sometimes be written for devices that have already been designed, such as the parts in a library, there may be situations in which this assumption is not valid. A simple example of this is the RS flip-flop. A common implementation of this device will behave unpredictably if both inputs are true. Thus a specification of such a device might not allow the event $r<true$ to occur if the value on the *s* port is true. This is an example of a *constrained* specification. Additional features are needed for the language to deal with such specifications. Indeed, the ability to deal with constraints has been given as one of the motivations for creating this new language instead of using an existing one. In Chapter 7, the topic of constraints and their description in SuperC will be discussed.

3.4   Summary

This Chapter has examined some of the issues associated with the specification task in a design and validation methodology. Specification is required to describe both the intended function of a device before it is implemented and the functions of the components that will be used to make an implementation. In either situation, a specification language should assist the designer by allowing concise, intuitive description, and it should if possible encourage the writing of descriptions that accurately reflect the behaviour of real hardware. A language must be sufficiently expressive to allow the description of any type of device in adequate detail, including its timing characteristics.

These issues were studied by considering the CIRCAL language. In its basic form, that language has several shortcomings, notably a lack of conciseness for many devices of only moderate complexity. A language, 'enhanced CIRCAL', was developed by considering these shortcomings and adding features to the language to overcome them while retaining the power of pure CIRCAL for manipulation of behavioural expressions. This will enable the language to be used for formal verification in subsequent chapters.

To write specifications effectively it is necessary to have not only a suitable language but also some associated specification techniques. A number of such techniques for use with enhanced CIRCAL were presented. Some of the techniques, such as constructive specification, reduce the effort of writing specifications, while others provide the means for describing a wide variety of timing phenomena. It is also important to ensure that descriptions fit in with the CIRCAL model of behaviour, which is event-based. To this end, techniques were developed to ensure that events which model changes of port values do in fact correspond to real changes of value. In Chapter 5, the importance of these techniques in validation will be demonstrated.

While enhanced CIRCAL provided a much more useful descriptive medium than pure CIRCAL, it was still somewhat difficult to use compared to many other languages. Thus, another language was presented which is intended to be more 'user-friendly' than even enhanced CIRCAL.

The language, called SuperC, removes the designer to an extent from the modelling details of CIRCAL and incorporates some of the specification techniques described above, enabling him to concentrate on the function of the device being described. As well as removing some of the tedium of writing specifications, the new language prevents the designer from falling into some of the traps that are present in pure CIRCAL, such as describing devices so that their inputs are occasionally 'locked'. The main shortcoming of the language is that, in order to carry out manipulation on the behavioural descriptions, they must first be translated into enhanced CIRCAL, and this translation is not generally reversible. However, the language would provide a useful front-end for a CIRCAL-based design system.

FOUR

DESIGN

In the subtasks of Figure 1.2, specifying is followed by partitioning and describing. The latter two tasks are generally quite difficult to separate and will be discussed in this Chapter under the single heading of 'Design'. In an ideal world, partitioning might be considered the purely structural operation of splitting the box at the higher level into several smaller, interconnected boxes, and describing would consist of associating behaviours with these smaller boxes. A more realistic view of a human designer would be that partitioning consists of splitting the larger box into smaller boxes whose behaviour is only informally defined, probably in the designer's head; the describe phase then makes the behaviour of the smaller boxes formal.

A primary concern of this book is the integration of design and validation within a hierarchical framework. In order to assist the integration of the two tasks, it is to be hoped that design techniques that assist the validation process could be developed. This Chapter will examine some possibilities for the achievement of this aim.

Another important theme is the role of languages in supporting a hierarchical design methodology. Since the design task consists largely of making choices, assistance in this task may be provided by reducing the number of choices available to a designer. The way in which a language may be used to represent this restriction will be presented.

A language may also assist the design phase by allowing a transformational approach to design to be adopted. In such an approach, the designer has a certain amount of freedom to make design decisions, but is forced to preserve the behaviour of the specification as he transforms it into an implementation, which may therefore be guaranteed to be correct. An example of a transformational approach to design is described in Section 4.2.1.

Of course, design is not always carried out purely by a human designer. A large number of design aids are available either to assist a human designer or to perform completely the design task in certain circumstances.

Some of these automated design techniques and their role in the proposed design and validation approach will also be discussed.

Without doubt the most general 'design system', however, is the human designer; all other tools and techniques are limited to specific areas of application (and may certainly outperform a human designer in these areas). Some of these tools, especially those referred to as 'expert systems', have been developed with the aim of mimicking the approach of a human designer. For these reasons it is appropriate to begin this chapter with an examination of the way in which a human designer might approach the design task in a hierarchical methodology.

4.1 Manual Design

It has been mentioned that design usually consists of partitioning and describing. For a human designer, partitioning often is effected by the drawing of circuit diagrams; these define the smaller boxes (components) from which the current box will be constructed and the interconnections between the components. That is, they provide an internal structural description of the device being implemented. In some design systems [Lattice 85, Morison 85], a designer is required to write this structural description linguistically. This is probably a less intuitive way to work, but is essentially equivalent, and it is not very difficult to translate between graphical and textual representations of circuit structure. Textual representations have the advantage of being both human- and machine-readable; this makes them suitable for input to simulators, for example. They also can be parameterised, which is not readily accomplished with graphical representations. Many languages for the formal representation of structure exist [Nash84]; the features of a number of languages for the description of circuit structure were discussed in Section 2.2.

The partitioning task can be, illustrated by considering again the full adder pictured in Figure 2.1. In designing this device, it may be initially considered as a box with inputs `X`, `Y` and `Cin` and outputs `Cout` and `Sum`. Figure 2.1 shows one possible way of partitioning the device into a pair of half adders and an Or gate.

Some designers would contend that a circuit diagram contains behavioural as well as structural information. This is true in a sense, in the same way that a conventional picture of an inverter, drawn as a triangle with a circle at one vertex, conveys behavioural information. Certain words and shapes *imply* certain behaviours, so that if one sees a box with 'RAM' written on it, one has a reasonable idea of its function. Such pictures do not, however, provide the *formal* description of behaviour that is required for a formal design and validation methodology; the structural information that they convey, on the other hand, is entirely adequate.

Having defined the internal structure of a device, with possibly some informal behavioural information as well, the partitioning phase must be followed by the description phase in which the behaviours of the components are formally defined. Sometimes to associate a behaviour formally with a box may require the writing of its specification, a task that is essentially identical to the specification of a system at the top level. For example, a specification of the half adder might be written at this stage. In other situations it may be possible to find the description of the required box in a library; in this case it would be expected that an implementation of that box also exists in the library. An Or gate, for example, is sufficiently common that it might be stored in a library. If the hierarchical design process is viewed as a tree, each node being a device and its children being the boxes from which it is constructed, then the design of a system is complete when all the leaves of the tree exist as implementations in the library.

Much of the discussion to date has assumed a rigorously top-down approach to design. In fact there are many situations where such an approach cannot or should not be taken. For example, attempts to implement a device may lead to a deliberate revision of its specification. Then, this revision may require that the devices to which the re-specified device is to be connected must also be re-specified; alternatively it may be necessary to formulate a new specification at a higher level in the hierarchy. Also, it is only by allowing some measure of bottom-up design that the available 'leaves' in the library (of standard cells or whatever) can be most efficiently used; for example, a library of gates could be used bottom up to design flip flops, which in turn could be used to design counters, etc. A knowledge of the availability of such parts would clearly influence the design decisions made at higher levels in the design hierarchy.

When carried out by a human designer, design may be viewed as a creative process. Koomen attempts to model the designer's creative input in terms of information theory [Koomen 85]; as the design progresses from an abstract specification to a concrete implementation, the amount of information must be increased with each step down the hierarchy, and this information is usually provided by the designer. Clearly a black box with an attached specification has less information than the same box after its internal structure has been defined and the behaviours of its component parts defined. The skill and experience of the designer will determine what information he chooses to add, i.e. how he chooses to partition a device into boxes and what behaviours he chooses to assign to them. It is this creative aspect of design which makes it so difficult to automate. In a later section some of the issues of the automation of design and how this fits in to a hierarchical design and validation methodology will be addressed.

While design and validation are often viewed as totally separate tasks, the case has already been made in previous chapters for the integration of the two tasks. This being the case, it is reasonable to ask whether the design task can be approached in such a way as to assist the verification task. The next Section addresses this question.

4.1.1 *Design for Validation*

One of the main arguments for hierarchical design is that, as a special case of the established technique of problem reduction, it enables a complex problem to be broken down into pieces of more manageable size. Within limits, a greater number of levels of hierarchy will result in a greater reduction in the complexity of the problem. There is, however, a *cost* associated with increasing numbers of levels, as specifications of components and their interconnections have to be written at every level. This cost must be traded off against the gains obtained by using more levels. In the integrated approach to design and validation proposed in the preceding chapters, validation takes place between adjacent levels in the hierarchy as soon as they have been designed. One way in which design can assist validation, then, is to use closely-spaced levels in the hierarchy to reduce the complexity of individual validation steps.

For this point to be demonstrated effectively, the nature of validation must be more fully explained, a task that will be tackled in the next Chapter. For the moment, it is sufficient to say that validation consists of demonstrating that an implementation satisfies a specification, and that this generally involves matching a complex behaviour (the implementation) with a simpler behaviour (the specification). The greater the 'gap' between these two levels, the more difficult it becomes to establish their equivalence. In the example presented in Chapter 6, the way in which design can be carried out to reduce this gap, and the consequent reduction in verification effort, is demonstrated.

4.1.2 *Restricting Design*

The primary argument that has been used to justify a language-based approach to design so far is that such an approach is essential to the integration of validation into the hierarchical design process. In addition to this, it seems reasonable to suppose that a language may be able to assist in the design task itself by *restricting* the designer. This situation may be thought of as the hardware analogy to high-level programming languages; these languages restrict the programmer's access to the full function of the computer, but in so doing may improve his productivity and increase the program's likelihood of being correct.

Silicon compilers derive their name by analogy to software compilation, so the input languages to these may be thought of as being analogous to high-level programming languages. This approach to design is discussed in more detail below. However, a design language may be able to restrict a designer without forcing him to depend on a specific design automation tool with its attendant disadvantages. A design language could be formulated so that a designer using it is restricted in a helpful way, with the extent of the restriction depending on the designer's level of expertise. The following example will illustrate how this might be done.

A frequently adopted approach to the design of large circuits involving combinational logic requires that globally clocked latches are placed between 'lumps' of combinational logic of a certain size. A circuit of this type is pictured in Figure 4.1. By choosing a clock period which is

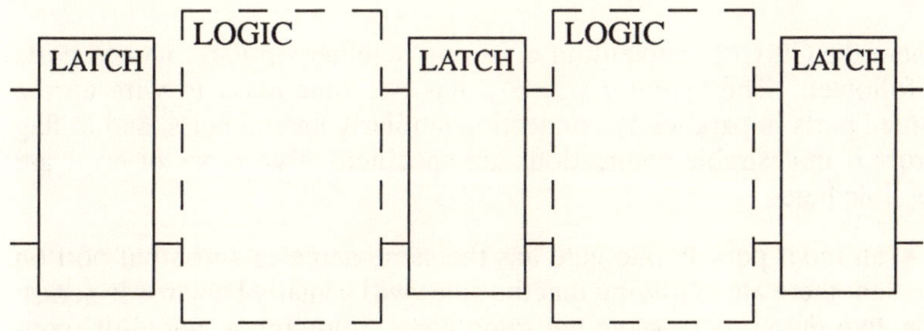

Figure 4.1: Latches in combinational logic design

longer than the delay across any one lump of logic, a system that will behave reliably can be constructed fairly easily. An experienced designer seeking to achieve maximum speed or minimum area might choose to ignore this design discipline, preferring to analyse the timing behaviour of each piece of logic in detail. A less experienced designer, however, would be considerably assisted by such a discipline, which ensures uniformity of delays across the circuit.

In order to provide assistance to a novice designer, high-level language constructs could be defined. In the above example, suitable high-level constructs would be a set of 'combinational wire-up' operators. For example, it might be decided to provide a 'wire-in-series' operator and a 'wire-in-parallel' operator. The arguments to the wire-in-parallel operator would just be gates chosen from some library of available parts. For example, in order to describe the configuration of gates pictured in Figure 4.2, one could write:

```
pblk1 = wire_p(and(a,b,c),and(b,f,e),inv(f,g))
```

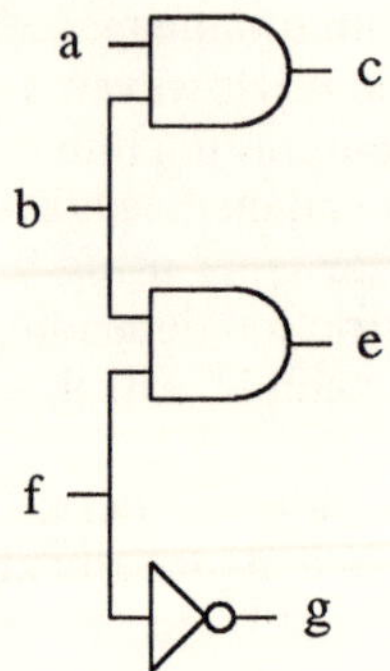

Figure 4.2: Gates wired in parallel

where the CIRCAL convention of wiring together similarly named ports is followed. The operator `wire_p` has two functions: to wire up the named parts in parallel by connecting similarly named ports, and to flag errors if undesirable connections are specified. Two types of error are possible here:

- an input port on one gate has the same name as an output port on another gate, implying that the gates will actually be wired in series;
- two output ports have the same name, implying a 'wired-or' configuration, the behaviour of which is not defined.

The actual wiring up can be achieved by translating the above description into CIRCAL, as:

```
pblk1 <= and * and[f/a][e/c] * inv
```

assuming that the definitions of `and` and `inv` had sorts $\{a,b,c\}$ and $\{f,g\}$ respectively.

The 'wire-in-series' operator would be more involved, as its function includes ensuring that latches are inserted between appropriate numbers of gates. If, for example, the delays of the available gates are such that three gates in series can be guaranteed to have a delay less than the clock period, then the wire-in-series operator could be defined such that it must have exactly three arguments, which may be either gates or parallel blocks. As well as wiring its arguments together, the series operator would wire a latch onto the end of the new block. Thus blocks defined with this operator could be safely wired together without the designer needing to be concerned about the delays of individual gates. An example circuit appears in Figure 4.3. Its description using the two high-level operators would be as follows:

```
pblk0 = wire_p(inv(p,a),nand(x,y,b))
pblk1 = wire_p(and(a,b,c),and(b,f,e),inv(f,g))

sblk  = wire_s(pblk0,pblk1,and3(c,e,g,z))
```

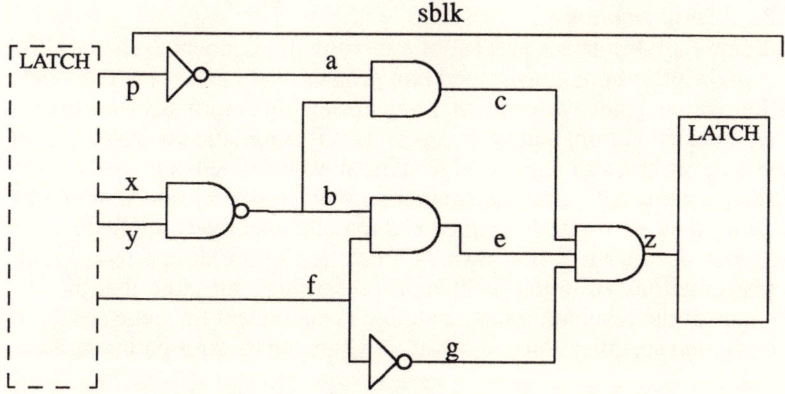

Figure 4.3: Gates wired in series and parallel

The latch shown as a solid box is automatically attached by the wire-in-series operator `wire_s`; the latch drawn with dashes would be from the previous stage.

This example suggests the following approach to language-based design: techniques that lead to safe designs, such as the one presented above, may be encapsulated in the design language. A novice designer may be given a very restricted set of operators with which to work; these will limit the range of design options open to him, thus guiding him in the design task and improving his chances of producing correct designs. A more experienced designer is given access to a wider range of operators, giving him greater flexibility and the capacity to make fuller use of the available technology, at the cost of more difficult design decisions and increased chance of error. In this simple example, only two levels of operator were presented: a novice designer would be allowed to use only the wire-in-parallel and wire-in-series operators, while an expert would be able to use the full range of CIRCAL structural operators. In a full design system, there would be perhaps several levels of operator corresponding to a variety of levels of expertise.

It is now clear that language features may directly assist the design task by restricting the designer in a useful way. A more realistic example than the one presented above is the gate-array design language

WISLAN [Vaidya 83a] which provides restricted access to the language Base CONLAN [Piloty 82]. Another way in which the designer may be restricted is by the need to satisfy contextual constraints, which are the subject of Chapter 7. A language may also provide design assistance by supporting design transformations. This topic is discussed below.

4.2 Transformation

In a sense all design is a process of transformation, in which abstract circuit descriptions are transformed into progressively more concrete ones. A human designer performs this transformation informally and manually; a design automation system performs it automatically according to some algorithm, although the algorithm may be defined only by the code of the system. Of particular interest in this Section are those transformations that are formally defined and may be selectively applied by the designer in order to work towards a solution to the design task. Such transformations are often 'behaviour-preserving', meaning that the behaviour of the resultant implementation is equivalent (in some sense) to the original specification behaviour to which the transformation was applied.

This approach to design represents a middle path between fully manual and fully automated design. It may still offer the guarantee of correctness and some of the reduction of design effort which come from automated techniques, but retains some of the flexibility and creativity of manual design. For this reason it has been the subject of considerable recent attention. Examples of transformational approaches to design include:

- a design system based on the LTS language that was described in Section 2.2.2 [Milne 86b];
- the work of Martin and others [Burns 88, Martin 86], which relies on the transformations of specifications written in an version of CSP [Hoare 78] to a form in which they can be readily synthesised as self-timed (i.e. unclocked) circuits;
- the approach of Sheeran[Sheeran 88] and others[Luk 88] which uses the RUBY descriptive framework[Sheeran 86] and relies on correctness-preserving transformations to improve the efficiency of designs.

An example of a simple behaviour-preserving transformation using CIR-CAL is presented below.

4.2.1 A CIRCAL-*based Transformation*

One way in which a correctness-preserving transformation could work is as follows: given a black box to be designed and its specification, the designer partitions the box into two components, using his own skill to decide what the sorts of the two components will be, and also to write a behavioural description of just one of the components. He may decide to extract that behaviour from a library. Then, an algorithm could be applied to determine the behaviour of the other component so that the two components wired together implement the specification. If the algorithm failed to find a suitable behaviour for the second component, the designer would need either to re-specify the other one or to choose a completely new partitioning. Thus the designer still has freedom in choosing the first component's behaviour, but is constrained by the algorithm to ensure that the implementation satisfies the specification. Such an approach represents a behaviour-preserving transformation which could be applied repeatedly until the specification was broken down into sufficiently small pieces.

In order to illustrate how this approach might be effected in CIRCAL, a slightly simplified problem will be tackled. Firstly, the implementation is restricted to have no hidden ports, i.e. the sort of the specification equals the union of the sorts of the two components that comprise the implementation. Secondly, all guards are required to include a tick of the universal clock; this will prevent the occurrence of events on one component that are completely independent of those on the other component.

The behaviours of the two components of the implementation can be written as

$$A = \sum a_i A_i \text{ and } B = \sum b_j B_j$$

and the specification as

$$C = \sum c_k C_k$$

The composition operator was defined in Section 2.3.3. With the restrictions described above, only the third term in the expansion of $A*B$ remains. This must equal the specification, so

$$\sum c_k C_k = \sum_{(b_j \cap M)=(a_i \cap N) \neq \phi} (a_i \cup b_j)(A_i * B_j)$$

where M is the sort of A and N the sort of B. Because of the restrictions in this simple example, the a_i and b_j can be easily found given the c_k, which are simply the guards of the specification. Since each c_k corresponds to the union of some a_i and b_j, the guards of the component behaviours can be found by removing those events that are not in their sorts. That is

$$a_i = c_k \cap M$$
$$b_j = c_k \cap N$$

It then follows that

$$C_k = A_i * B_j$$

so that the procedure for finding the behaviours A_i and B_j is exactly the same as that just demonstrated.

A simple example illustrates how this transformation could be used. A device is specified as

```
C  <= {a, t}C1 + {b, t}C2 + {c, t}C3
    + {a, b, t}C4 + {a, b, c, t}C5
C1 <= ...
```

The designer decides to partition the box into a component A of sort $\{a,b,t\}$ and a component B of sort $\{b,c,t\}$. $a_1 = c_1 \cap M = \{a, t\}$. $b_1 = c_1 \cap N = \{t\}$. Continuing this procedure for all five guards results in

```
A <= {a, t}A1 + {b, t}A2 + {t}A3 + {a, b, t}A4
B <= {t}B1 + {b, t}B2 + {c, t}B3 + {b, c, t}B4
```

The behaviours `A1-4` and `B1-4` would need to be determined from the definitions of `C1-5`. It can easily be shown that the behaviour of A*B can perform all the actions of C. In this case, it can also perform several additional actions, but this is acceptable, as will be shown in the following Chapter; such an implementation is said to *satisfy* its specification.

This particular transformation is really very limited in usefulness. It has not been defined for enhanced CIRCAL, the language in which most useful descriptions must be written. The restriction that guards must contain ticks is acceptable, but the requirement that the implementation have no hidden ports limits the application of this approach severely. However, it serves to illustrate some important points: that a language which supports formal reasoning about behaviour can also support a transformational approach to design; and that such an approach can leave a designer with some creative freedom while still giving him assistance in the design task and ensuring that design steps are correct.

A general point to be made here is that the more assistance a designer is given, the less flexibility he has. The approach that gives a designer the most assistance and limits his flexibility to the greatest extent is automated design, which is the subject of the next Section.

4.3 Automated Design

The field of Design Automation (DA) is a vast one and it is not appropriate to attempt to cover it in detail here. What is important here is to examine how design automation tools, such as silicon compilers, may be fitted in to the proposed approach to design and how this affects the specification and validation tasks. It is also interesting to study whether this approach, which has been motivated mainly by the desire for hierarchical validation, has anything to contribute in terms of design automation.

The ideal design automation system would accept some very abstract specification of the required device and from that would produce, without any human intervention, a complete implementation which was guaranteed to perform the required task correctly. Furthermore, the implementation would be at least as efficient as anything a human designer could produce. Such a system would completely remove the need for a design and validation methodology — all that would be required to design perfect chips would be some means of extracting correct specifications to feed to the system. Unfortunately, such a system does not exist; current design automation tools are limited in their range of application, often produce less efficient designs than those produced manually, and are not always guaranteed to produce correct implementations. The fact that the systems are limited in scope means that they can only be used in some parts of the design process, and in order to preserve the guaranteed correctness that is the aim of the proposed methodology, it is necessary either to validate the output of DA tools against the input specifications or to validate the tools themselves.

4.3.1 Specification and Design Automation

If a design automation tool is to be used to design just part of a system which, as a whole, is being designed in the hierarchical manner proposed in Chapter 1, then the specification of that part has three important roles to play. Obviously it must convey sufficient information to the DA tool to enable it to produce an implementation. The specification must also be used to enable the validation of the *preceding* design step. And finally it may have to be used to permit the validation of the output of the design automation tool, if this tool has not itself been validated (i.e. if it has not been proven that the tool always produces correct implementations). The second of these roles dictates that the language that is accepted by the tool as input be the same specification language that is in use for the methodology. If the output of the tool is to be validated, then it is important that the output can at least be translated into a form which will permit this validation to take place. Preferably, the output should be in the same language as the input to the tool, but this may not always be possible, as discussed below.

One question that arises is whether these three roles for a specification being used for design automation place contradictory requirements on the specification language. Is it reasonable to expect, for example, that a single language should be suitable both as input to a PLA generator (a very simple design automation tool) and as a specification language supporting formal reasoning about hardware behaviour? Certainly there have been relatively few attempts to unify languages in this way, with many automated design tools requiring input specifications in their own purpose-built language [Lattice 85, Siskind 82]. In recent years, however, with the numerous attempts to arrive at a standard hardware description language [Morison 85, Piloty 82, USAF 84], the need to develop languages that are suitable for widely disparate tasks has been recognised and catered to. The criteria by which languages were selected for study in Chapter 2, especially the ability to describe a wide range of hardware, should ensure that they are suitable for all of the above roles.

4.3.2 Validation of Automated Designs

The validation of the output of an automated design system is just like any other validation if the output is produced in a form which supports validation. If the output of the system is in some less convenient form, such as the mask information that might be output by a PLA generator, then the validation task becomes a little more complicated. If validation is to be performed by simulation, then the solution is to use a simulator that can accept mask information, or at least a circuit description extracted from the mask information, as input. The simulated behaviour of the implementation can then be compared with the behaviour that was specified, possibly by also simulating the specification with the appropriate simulator. If formal methods are to be used, then a model of the implementation's behaviour must be derived in the mathematical framework that is to be used for the verification. This may be done by circuit extraction and behavioral description of the resultant circuit.

An alternative approach to the validation of devices produced by DA tools has already been alluded to. This is to validate the tool itself, thus guaranteeing that the implementations it produces will always satisfy the specifications that are fed to it. A discussion of this approach will be postponed until the issues of validation are discussed in more detail in Chapter 5.

4.4 Summary

In this Chapter, the design phase, consisting of partitioning and description, has been examined. As the aim is to integrate design and validation, an important consideration is the way in which certain approaches to design affect the validation task. In particular, the importance of the distance between levels of hierarchy in determining the difficulty of validation was discussed, and this point will be returned to in Chapter 6.

Design is an essentially creative task, often relying heavily on a designer's skill. While it is widely accepted that hierarchical design reduces the difficulty of this task, this Chapter has gone some way toward showing that a language-based approach to design can also provide the designer with considerable assistance.

By analogy to high-level programming, a language for hardware design may also be high-level in that it offers the designer access to more powerful constructs and prevents him from committing certain errors. Such a language may provide assistance to a designer by reducing the options open to him in a given situation. The idea of levels of language which restrict a designer according to his level of expertise was presented and illustrated by the well-known example of clocked latches placed between lumps of combinational logic.

A language which supports formal manipulation of behaviours may be able to provide design assistance by supporting a transformational approach to design. Transformations which preserve the behaviour of a specification, thus ensuring a correct implementation, while allowing a designer to make some creative input, were introduced. A simple example of such a 'behaviour-preserving' transformation using pure CIRCAL was presented.

There is one further way, not discussed in this Chapter, in which languages may assist design. It is only in language-based design that constraints may be reasoned about formally, and this may be of considerable importance. This topic is dealt with in Chapter 7.

Design cannot be fully discussed without some treatment of the area of design automation. Design automation tools, while suffering from some notable shortcomings, are often able to produce designs quickly and reliably. Such tools, when used in a hierarchical design methodology, place certain requirements on the specification language; fortunately these requirements are quite close to those that were outlined in Chapter 3. Either the output of the tools, or the tools themselves, need to be validated, as will be discussed in the next Chapter. These issues serve to highlight the interdependence of the features of a design language and the performance of the tasks of specification, design and validation.

FIVE

VALIDATION

In Chapter 1 the motivation for a hierarchical approach to design and validation was given: in essence, this was to enable the detection of design errors at the earliest possible point in the design process, thus reducing the cost of such errors. The tasks in the methodology that have been described in the preceding chapters are essential to it, but it is on the validation task that the whole process hinges, for it is by performing this task that design errors can be detected, assuming that the design task has not been carried out in such a way as to guarantee their exclusion.

As in the previous Chapter, the task described here splits into two parts. In Figure 1.2 these were shown as 'Compose' and 'Compare'. Because they are virtually never carried out in isolation, and the way in which one is performed strongly affects the other, they are discussed here under the single heading of 'Validation'.

Broadly speaking, there are two approaches to the validation task. The more widely used of these is simulation: the attempt to establish the behaviour of an implementation by calculating its response to a selection of input stimuli. The first part of this Chapter will deal with this method of validation. The principal shortcoming of simulation, its failure to prove the absence of design errors except by the unacceptably costly technique of exhaustive simulation, will be discussed and a possible way to overcome this problem will be examined.

The other main approach to validation is referred to as verification. This word is often taken to mean different things by different people, but the definition used here will be the formal, mathematical proof that an implementation behaves as is required by its specification. This technique avoids the main shortcoming of simulation, offering a guarantee that an implementation behaves correctly. The cost of this certainty is the complexity that is encountered in the use of formal methods. Ways of tackling this complexity will be discussed.

A theme throughout this Chapter will be the requirements that a particular validation technique places on the specification language; once again CIRCAL will be used as the main example, and its suitability for both simulation and formal verification will be examined. Also, by attempting to validate devices specified using the techniques of Chapter 3, the effectiveness of those techniques will be evaluated.

5.1 Simulation

The simulation of an implementation consists essentially of predicting or calculating the responses it will give to a range of input stimuli. In Sections 5.1.2 and 5.1.3, two different approaches to this task are presented. These are followed by some examples and a discussion of the problems that may arise in simulation, including difficulties caused by certain specification techniques. Finally, a way of improving the confidence gained from simulation is discussed. Before performing a simulation, however, it is necessary to specify the input stimuli to be applied. This topic is discussed in the next Section.

5.1.1 Input Specification

If the behaviour of a device under certain input stimuli is to be established, then there must be some way of describing those input stimuli. It is common for a simulator to be equipped with a specialised language for this purpose, but in fact it should be possible to use a general-purpose hardware description language. In CIRCAL a sequence of input stimuli can be represented by a series of guards, as the following example illustrates.

Example

Suppose that in order to simulate a counter it is required to load it with the value 15, then apply 5 successive clock pulses. These input stimuli could be specified in CIRCAL as follows:

```
START <= {data<15}{asl<true}CLOCK(5)
CLOCK(n) <=
    if noteq(n,1) then
        {clk<true}{clk<false}CLOCK(n-1)
  + if eqs(n,1) then
        {clk<true}{clk<false}  /\
```

Explanation

The counter has an asynchronous load port (*asl*) which, when triggered by a rising edge, causes the value on the *data* port to be loaded. The stimulus generating device described above therefore begins by putting the value 15 on the input port *data* before providing the rising edge on *asl*. It then moves to state CLOCK(5), from which state pulses are generated on the *clk* port until its state variable is decremented to 1, at which time a final pulse is generated and the device terminates.

5.1.2 Conventional Simulation

To simulate a device that is specified behaviourally (i.e. with no reference to its internal structure) is not difficult. Whatever language it is described in, it should be possible to predict what outputs it will generate given a certain pattern of inputs. The event-based model of CIRCAL makes it quite a suitable language for this task, and the response of a device to input stimuli can be easily determined. It should be emphasised that the simulation performed here does not depend on many of the unusual features of CIRCAL; virtually any hardware description language could be used in this role.

Example

The loadable counter mentioned above could be described in CIRCAL using only behavioural operators in the following way:

```
COUNT(false,false,n,p) <=
   {clk<true,out<incr(p)}
                 COUNT(true,false,n,incr(p))
 + {data>x:noteq(x,p)}COUNT(false,false,x,p)
 + {asl<true,out<n}COUNT(false,true,n,n)
COUNT(false,true,n,p) <=
   {clk<true,out<incr(p)}
                 COUNT(true,true,n,incr(p))
 + {data>x:noteq(x,p)}COUNT(false,true,x,p)
 + {asl<false}COUNT(false,false,n,p)
COUNT(true,false,n,p) <=
   {clk<false}COUNT(false,false,n,p)
 + {data>x:noteq(x,p)}COUNT(true,false,x,p)
 + {asl<true,out<n}COUNT(true,true,n,n)
COUNT(true,true,n,p) <=
   {clk<false}COUNT(false,true,n,p)
 + {data>x:noteq(x,p)}COUNT(true,true,x,p)
 + {asl<false}COUNT(true,false,n,p)
```

Explanation

The four state parameters of this device represent the values on the ports *clk* (the clock), *asl* (asynchronous load), *data* (input for loading a value) and *out* (the output port). The first two are of type boolean, the second two are *n*-bit integers. The function `incr(p)` should be defined to add 1 to p modulo 2^n. In the states where the value on *clk* is false (the first two) an event `clk<true` causes the output to increment. In the states where the value on *asl* is false (the first and third) an event `asl<true` causes the output to be set to the value currently on the data input port *data*. All other events that may occur simply cause new values to be placed on input ports.

In order to carry out a simulation, a start state must be calculated or chosen. Taking this to be the first of the above four states, then it is quite clear that from this state the events generated by the input device described above, `{data<15}` `{asl<true}` will lead the counter into state `COUNT(false, true, 15, 15)` and that the following 5 rising edges on the *clk* line will cause the expected incrementing of the output.

Simulation of a purely behavioural description in this way could be useful in checking that a large specification correctly reflects the intentions of the designer. It could also be used to provide a useful description of the behaviour of a specification against which the simulation results for the implementation could be compared. The obtaining of these results requires that a description that is not purely behavioural be simulated, and this is a less straightforward task.

In many simulators, the approach that is taken to establish the behaviour of a constructed device is to carry out simulations of each of the component devices concurrently, treating each component in a manner similar to that just described. The following simple example will illustrate how this might be done.

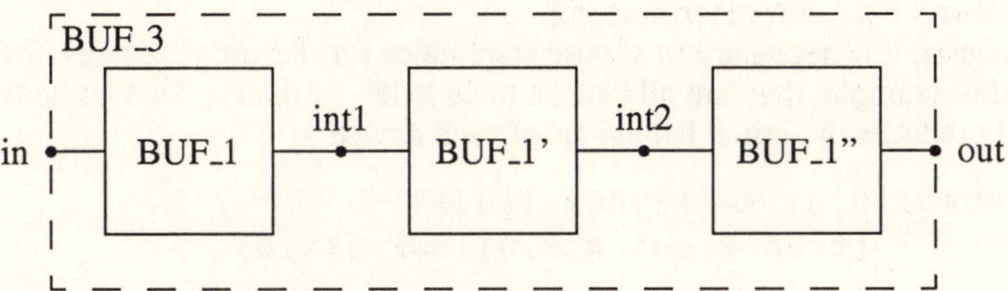

Figure 5.1: Construction of 3 unit buffer

Example

The constructed device to be simulated, pictured in Figure 5.1, is a 3 time-unit delay buffer, constructed from three single unit delay buffers. The behaviour of a unit delay buffer is

```
BUF_1(p,q) <=
    if eqs(p,q) then ({t}BUF_1(p,p)
        + {t,in>x:noteq(x,p)}BUF_1(x,p))
  + if noteq(p,q) then ({t,int1<p}BUF_1(p,p)
        + {t,in>x:noteq(x,p),int1<p}BUF_1(x,p))
```

and the other two buffers have identical behaviours, suitably relabelled, as follows:

```
BUF_1'  <= BUF_1 [int2/int1][int1/in]
BUF_1"  <= BUF_1 [out/int1][int2/in]
```

Explanation

The buffer has a value p stored on the input *in*, and a value q on the output *int1*. If these are equal, then a tick of the universal clock will produce no change on the output. If they are unequal, then the value p will appear on the output when a tick occurs. In either case, the tick may also be accompanied by the arrival of a new value on the input port.

Now suppose that the input stimulus is described by the following device:

```
STIM <=
  {in<5,t}{t}{t}{t}{in<2,t}{in<7,t}{t}{t}{t}/\
```

Explanation

This device places the value 5 on the input port, then waits for 3 ticks, then places the values 2 and 7 on the input on successive ticks, then waits another 3 ticks before terminating.

Again, it is necessary to choose start states for the simulated devices; in this example, they are all chosen to be BUF_1(0,0). This immediately reduces the initial behaviour of each device to

```
BUF_1(0,0) <= {t}BUF_1(0,0)
        + {t,in>x:noteq(x,0)}BUF_1(x,0)
```

To establish the events that will take place it is helpful to tabulate the possible events for each device.

Simulation	Device			
Step	STIM	BUF_1	BUF_1'	BUF_1"
1	{in<5,t}	{t} {in>x,t}	{t} {int1>x,t}	{t} {int2>x,t}

Note that the predicates on input parameters have been omitted for reasons of space only.

Given that an event can only take place on a port if all devices connected to that port are able to accept an event on it, then it can readily be established from the above table that only one guard is possible, i.e. {`in<5,t`}. This is because all devices can accept the event `t` and both STIM and BUF_1 can accept `in<5`. The event `int1>x`, for example, is not possible because it cannot be accepted by BUF_1. With the occurrence of {`in<5,t`}, BUF_1 moves into the state `BUF_1(5,0)`. The table can now be updated and the next event calculated, and so on until the simulation ends or is terminated. The result of this process is tabulated in Table 5.1.

Step	Device				Result
	STIM	BUF_1	BUF_1'	BUF_1"	
1	{in<5,t}	{t}	{t}	{t}	{in<5,t}
		{in>x,t}	{int1>x,t}	{int2>x,t}	
2	{t}	{t,int1<5}	{t}	{t}	{t,int1<5}
		{in>x,t,int1<5}	{int1>x,t}	{int2>x,t}	
3	{t}	{t}	{t,int2<5}	{t}	{t,int2<5}
		{in>x,t}	{int1>x,t,int2<5}	{int2>x,t}	
4	{t}	{t}	{t}	{t,out<5}	{t,out<5}
		{in>x,t}	{int1>x,t}	{int2>x,t,out<5}	
5	{in<2,t}	{t}	{t}	{t}	{in<2,t}
		{in>x,t}	{int1>x,t}	{int2>x,t}	
6	{in<7,t}	{t,int1<2}	{t}	{t}	{in<7, int1<2,t}
		{in>x,t,int1<2}	{int1>x,t}	{int2>x,t}	
7	{t}	{t,int1<7}	{t,int2<2}	{t}	{int1<7, int2<2,t}
		{in>x,t,int1<7}	{int1>x,t,int2<2}	{int2>x,t}	
8	{t}	{t}	{t,int2<7}	{t,out<2}	{int2<7,out<2,t}
		{in>x,t}	{int1>x,t,int2<7}	{int2>x,t,out<2}	
9	{t}	{t}	{t}	{t,out<7}	{t,out<7}
		{in>x,t}	{int1>x,t}	{int2>x,t,out<7}	

Table 5.1: Possible and resultant events for BUF_3

From the above it is quite clear that events on the output lag those on the input by three time units. For a more complicated device, it would be preferable also to simulate its behavioural specification and compare the results of that simulation with those obtained from the simulation of the constructed device.

It is noteworthy that, in this method of simulation, no use was made of the feature of CIRCAL that enables behaviours of constructed devices to be established by manipulation of the behavioural expressions corresponding to its component parts. This method of simulation would work equally well if the language used to describe the components had no such feature. If a single device described in the language can be simulated just as the loadable counter was simulated at the start of this Section, then a circuit constructed from devices described in that language can also be simulated. It is for this reason that many hardware description languages do not offer facilities to construct behaviours mathematically, since simulation can be performed without that capability.

The manner in which behavioural descriptions of components may be manipulated in CIRCAL offers a rather different approach to the simulation task from that described above. This approach, which was proposed by Milne [Milne 85b] and described more fully by Traub [Traub 86], warrants some discussion here.

5.1.3   *Simulation by Manipulation*

In the example just described, the constructed system could be described structurally as

```
BUF_3 <=
    BUF_1 * BUF_1[int2/int1][int1/in]
          * BUF_2[out/int1][int2/in]
```

The composition (*) and relabelling ([/]) operators convey essentially structural information, but have a formal behavioural interpretation. If the start states for the three buffers are as before then an expansion of the above expression using the definition of the composition operator yields

```
BUF_3 <=
   {t}(BUF_1(0,0)*BUF_1'(0,0)*BUF_1"(0,0))
 + {t,in>x:noteq(x,0)}
      (BUF_1(x,0)*BUF_1'(0,0)*BUF_1"(0,0))
```

and subsequent composition of this with the expression for STIM yields

```
 BUF_3 * STIM <=
   ({t}(BUF_1(0,0)*BUF_1'(0,0)*BUF_1"(0,0))
  + {t, in>x:noteq(x,0)}
      (BUF_1(x,0)*BUF_1'(0,0)*BUF_1"(0,0)))
  * ({in<5,t}{t}{t}{t}{in<2,t}{in<7,t}
                            {t}{t}{t}/\)
 = {in<5,t}(BUF_1(x,0)*BUF_1'(0,0)*BUF_1"(0,0))
   * ({t}{t}{t}{in<2,t}{in<7,t}{t}{t}{t}/\)
```

Hence, the composition operator provides the result that the only possible event is {in<5, t}, just as before. Repeated applications of the composition operator could be used to establish the sequence of events that would follow. The advantage of using CIRCAL as the input specification language is clearly seen here, as this enables the simulation to be carried out entirely by expansion of the composition operator.

It may be thought that the simulation method just described is not greatly different from the method of the preceding Section. In a sense this is true, as the use of the composition operator really just formalises the process of establishing the resultant event when each device has a number of options open. The key difference, however, is that the first method could be used with a language which did not have the formal semantics that enable construction of behaviours to be performed; such semantics are crucial to the second method. It is also noteworthy that in attempting to implement a simulator of CIRCAL expressions, Johnson [Johnson 86] found that this second method was preferable for ease of implementation.

5.1.4 *Problems*

Section 3.2 introduced a number of specification techniques for use with the CIRCAL language that either eased the specification task or encouraged the writing of more useful or accurate specifications. Many of these techniques evolved from the process of trying to make use of specifications for simulation and verification. This Section discusses some of the problems that may arise from unsuitable specification techniques and how these may be avoided.

Unforced Outputs

In Section 2.3.2 a description similar to the following was given for an inverter with delay:

```
INV(0,1) <= {in<1}INV(1,1)
INV(1,1) <= {out<0}INV(1,0)
   + {in<0,out<0}INV(0,0) + {in<0}INV(0,1)
INV(1,0) <= {in<0}INV(0,0)
INV(0,0) <= {out<1}INV(0,1)
   + {in<1,out<1}INV(1,1) + {in<1}INV(1,0)
```

The intention of this specification is to show that INV exhibits a propagation delay without attaching a value to that delay.

If this device is simulated under the following input stimuli

```
STIM <= {in<1}{in<0}{in<1}/\
```

then the first event to take place, given that the start state is INV(0,1), may be determined by an expansion of the composition operator as follows:

```
STIM * INV(0,1) <=
    ({in<1}{in<0}{in<1}/\) * ({in<1}INV(1,1))
 = {in<1} (({in<0}{in<1}/\) * INV(1,1))
```

So, the first event to take place is `in<1`. To determine the next event, the required expansion is:

```
({in<0}{in<1}/\) * INV(1,1)
 = ({in<0}{in<1}/\) *
   ({out<0}INV(1,0) + {in<0,out<0}INV(0,0)
    + {in<0}INV(0,1))
 =   {out<0} (({in<0}{in<1}/\) * INV(1,0))
  + {in<0,out<0} (({in<1}/\) * INV(0,0))
  + {in<0} (({in<1}/\) * INV(0,1))
```

Rather than revealing exactly what the next events will be, this has given a choice of three guards. This is not what is generally expected of a simulator, which should describe the output behaviour for a given pattern of inputs. The fault here is not in the simulator but in the technique used to specify the inverter delay. The ability to specify a device such that it has a non-zero delay, without needing to specify the length of the delay, may seem quite attractive, but this example shows the problems that such an approach can cause.

In order to achieve a specification in which the delay is non-zero but otherwise unspecified, a more useful approach would be to use the method involving ticks of an external clock but leave the time interval between the ticks unspecified. The inverter could now be written as

```
INV(0,1) <= {in<1,t}INV(1,1) + {t}INV(0,1)
INV(1,1) <=
   {out<0,t}INV(1,0) + {t,in<0,out<0}INV(0,0)
INV(1,0) <= {in<0,t}INV(0,0) + {t}INV(1,0)
INV(0,0) <=
   {out<1,t}INV(0,1) + {t,in<1,out<1}INV(1,1)
```

The key difference between this specification and the initial one is that input and output events are now assumed to occur synchronously with ticks of the universal clock; the delay between a change of input value and a corresponding output change is now T, the time between successive ticks of the clock. This effectively 'forces' a change to occur on the output following an input change.

The stimulus generator STIM must now also be written with reference to the ticks:

```
STIM <= {in<1,t}{in<0,t}{t}{in<1,t}{t}/\
```

Repeated expansion of the composition operator on `STIM * INV(0,1)` now gives a unique sequence of output events:

```
STIM * INV(0,1) = {in<1,t}{in<0,out<0,t}
                  {out<1,t}{in<1,t}{out<0,t}/\
```

It should be noted that, although the delay of the inverter is now specified by T, no firm statement has been made as to how long T is; thus the goal of specifying only that delay is non-zero has been achieved, without the undesirable consequences of the first approach. This technique has potentially quite a wide range of application. It enables a specification to be written in such a way that the value output by a device is specified without an exact specification of how long it will take to calculate that value. This may be useful, for example, in the high-level specification of an arithmetic unit, when what is important is that it performs the appropriate functions, rather than the time taken to perform these functions. An elaboration on this technique is used in the major verification example of Chapter 6.

Fictitious Events

One of the specification techniques proposed in Section 3.2 was designed to ensure that only 'genuine' events could occur. That is, if an event is being used to model a change of value on a port, then only those events that represent actual changes should appear in a device description. An event such as `out<3` should not occur if the value on the port *out* is already 3. The consequences of writing specifications in which such fictitious events occur can involve a serious penalty in simulation efficiency, as the following example demonstrates.

Example

The single delay buffer of the previous Section could be described more concisely without regard for the validity of events as follows:

```
BUF_1(p)  <= {t,int1<p}BUF_1(p)
          + {t,in>x,int1<p}BUF_1(x)
```

Explanation

This device outputs the stored value p on the port *int1* with every tick of the universal clock, regardless of whether it is different from that already there. A value x may also be placed on the input simultaneously with a tick, and this value may or may not differ from the current value p.

The table to determine events would then be

Step	Device				Result
	STIM	BUF_1	BUF_1'	BUF_1"	
1	{in<5,t}	{t,int1<0}	{t,int2<0}	{t,out<0}	{in<5,t,int1<0
		{in>x,t,int1<0}	{int1>x,t,int2<0}	{int2>x,t,out<0}	int2<0,out<0}

Note that the resultant guard now involves an event on every port, compared with just two events in the previous case. In a complicated design, this could lead to a proliferation of events throughout the circuit, drastically increasing the time required to calculate the events taking place at each instant in time. It is for this reason that true event-driven simulation is a popular approach, as adopted for example in the VHDL project [Gilman 86, Lowenstein 86].

The aspect of simulation that is widely acknowledged as its most serious shortcoming is its failure to guarantee the absence of design errors. Errors can be shown to be present by simulation, and confidence in a design's correctness can be increased when several simulations have failed to reveal any errors. However, only if all possible combinations of device states and input stimuli are considered can the correctness of the device be guaranteed. This approach, called exhaustive simulation, is far too time-consuming to be considered seriously in the majority of cases, and so other methods of providing greater confidence in a design have been developed. The main one of these, formal verification, is the subject of the second part of this Chapter; the following Section discusses some approaches that lie between verification and conventional simulation.

5.1.5 *Proof by Simulation*

The task of verifying a k-definite system by simulation has been addressed by Bryant [Bryant 86]. Such a system has a behaviour such that any sequence of inputs of length k will place the system in a unique state. This class of systems includes finite state machines, but excludes some simple circuits such as registers. Such a circuit can be proven correct by setting all internal states to an undefined value 'X', then calculating the output after all possible input sequences of length k have been applied. This is simply exhaustive simulation. However, the unacceptable complexity of the task can be reduced if some values in the input sequences are unimportant to the output (such as the last n inputs to an $n + 1$ bit long shift register). In such a circumstance, values in the input sequences can be replaced by the undefined value 'X', thereby greatly reducing the number of sequences that must be tested to prove the circuit's correctness. There are a significant number of situations in which the savings offered by this approach are considerable.

A further simulation technique which may lead to guaranteed correctness is symbolic simulation [Carter 79, Darringer 79], in which the input stimuli are represented by variables rather than constant values such as true and false. Such a technique can readily be adopted using CIRCAL, as was done by Traub [Traub 86], as it is easy to specify an input stimulus generator which supplies variables of any type, and the techniques of Section 5.1.3 may then be applied.

5.2 Verification

This Section discusses some of the issues involved in mathematical proof of circuit correctness or verification. Most of the discussion will be of verification using CIRCAL, although some other approaches to the task will also be briefly introduced. As in the previous Section, the use of certain specification techniques and the avoidance of others will be justified by considering the consequences of applying them to the solution of actual problems. Some of the advantages and disadvantages of using CIRCAL for verification will be discussed, along with the advantages of verification over simulation. In order to illustrate how a simple verification proof might proceed a small example will first be presented.

5.2.1 A Simple Proof

The device to be constructed and verified is a buffer with a delay of two ticks. Its specification is as follows:

```
BUF(w,x,y) <= if eqs(x,y) then
    ({t}BUF(w,w,x)
    + {t,in>q:noteq(w,q)}BUF(q,w,x))
  + if noteq(x,y) then
    ({t,out<x}BUF(w,w,x)
    + {t,in>q:noteq(w,q),out<x}BUF(q,w,x))
```

Explanation

To model the delay of two ticks, a queue of values represented by the parameters w, x and y is used. If x and y are equal, then a tick produces no output change, but the values move one place along the queue. If x and y differ, then x is placed on the output port when the tick occurs. In either case the tick may be accompanied by the placing of a new value q on the input.

The buffer is to be implemented by wiring two inverters in series. This implementation is structurally described as

```
IMP <= INV[mid/out] * INV [mid/in] - mid
```

where the behavioural description of INV is

```
INV(m,n) <= if noteq(m,n) then
   ({t}INV(m,n)
   + {t,in>p:noteq(p,m)}INV(p,m))
 + if eqs(m,n) then
   ({t,out<not(m)}INV(m,not(m))
   + {t,in>p:noteq(p,m),out<not(m)}
       INV(p,not(m)))
```

Explanation

The parameter m represents the value on the input port, while n represents the value on the output. If these are unequal, the inverter is stable and the occurrence of a tick will cause no output change, but may be accompanied by an input event. If they are equal, the output must change on the next tick to the complement of its current value.

The verification proof consists of attempting to show that IMP 'is equivalent to' BUF. The definition of equivalence is discussed in more detail below. To begin the proof, it is helpful to use the fact that both INV devices are connected to the port *mid*. So, if the parameters of the second device are changed to c and d, then c must equal n. This assumption is not essential to the proof, but reduces the amount of manipulation required here. The proof can now proceed by applying the definition of the composition operator, given in Section 2.3.3, to obtain the following:

```
IMP <= INV[mid/out] * INV [mid/in] - mid
=   (if noteq(m,n) then ({t}INV(m,n)
      + {t,in>p:noteq(p,m)}INV(p,n))
   + if eqs(m,n) then
        ({t,mid<not(m)}INV(m,not(m))
      + {t,in>p:noteq(p,m),mid<not(m)}
          INV(p,not(m))))
* (if noteq(n,d) then ({t}INV(n,d)
     + {t,mid>p:noteq(p,n)}INV(p,d))
   + if eqs(n,d) then
        ({t,out<not(n)}INV(n,not(n))
      + {t,mid>p:noteq(p,n),out<not(n)}
          INV(p,not(n))))
- mid
= if and(noteq(m,n),noteq(n,d)) then
   ({t}INV(m,n)*INV(n,d)
   + {t,in>p:noteq(p,m)}INV(p,n)*INV(n,d))
+ if and(noteq(m,n),eqs(n,d)) then
   ({t, out<not(n)}INV(m,n)*INV(n,not(n))
    + {t,in>p:noteq(p,m),out<not(n)}
```

```
                 INV(p,n)*INV(n,not(n)))
  + if and(noteq(not(m),n),and(eqs(m,n),
                   noteq(n,d))) then
    ({t,mid<not(m)}INV(m,not(m))*INV(not(m),d)
      + {t,in>p:noteq(p,m),mid<not(m)}
          INV(p,not(m))*INV(not(m),d))
  + if and(noteq(not(m),n),and(eqs(n,d),
                   eqs(m,n))) then
    ({t,mid<not(m),out<not(n)}
        INV(m,not(m))*INV(not(m),not(n))
      + {t,in>p:noteq(p,m),mid<not(m),out<not(n)}
          INV(p,not(m))*INV(not(m),not(n)))
  - mid
```

Predicates can be removed or simplified; for example

```
and(noteq(not(m),n),and(eqs(m,n),noteq(n,d)))}
```

becomes

```
and(eqs(m,n),noteq(n,d))}.
```

The expression `IMP(m,n,d)` can be used to represent `INV(m,n)` `*` `INV(not(n),d)`. Removing events on the port *mid* and using the expression `IMP(m,n,d)` to represent `INV(m,n)*INV(not(n),d)` leads to

```
IMP(m,n,d)  <=
    if and(noteq(m,n),noteq(n,d)) then
      ({t}IMP(m,n,d)
        + {t,in>p:noteq(p,m)}IMP(p,n,d))
  + if and(noteq(m,n),eqs(n,d)) then
      ({t,out<not(n)}IMP(m,n,not(n))
        + {t,in>p:noteq(p,m),out<not(n)}
            IMP(p,n,not(n)))
  + if and(eqs(m,n),noteq(n,d)) then
      ({t}IMP(m,not(m),d)
        + {t,in>p:noteq(p,m)}IMP(p,not(m),d))
  + if and(eqs(n,d),eqs(m,n)) then
      ({t,out<not(n)}IMP(m,not(m),not(n))
        + {t,in>p:noteq(p,m),out<not(n)}
            IMP(p,not(m),not(n)))
```

This can be further simplified, using the fact that whenever `eqs(m,n)` is true, the second of `IMP`'s parameters becomes `not(m)`, whereas when `noteq(m,n)` is true, it becomes n, which must also equal `not(m)`. This is the only effect of these predicates, so they can be removed, reducing the expression to:

```
IMP(m,n,d) <=
    if noteq(n,d) then ({t}IMP(m,not(m),d)
        + {t,in>p:noteq(p,m)}IMP(p,not(m),d))
  + if eqs(n,d) then
        ({t,out<not(n)}IMP(m,not(m),not(n))
        + {t,in>p:noteq(p,m),out<not(n)}
            IMP(p,not(m),not(n)))
```

This will be able to perform the same initial actions as BUF if m = w, `not(n)` = x and d = y. After performing the initial actions, IMP will be able to perform the same actions as BUF if these same relationships hold between the new parameters of BUF and those of IMP. That is, m = w, `not(not(m))` = w, `not(n)` = x and d = x. The first three equations follow from the initial conditions on the parameters; the last equation needs to be true only if `eqs(x,y)` is true, in which case it follows from the initial assumption that d = y. It therefore follows that, if IMP and BUF commence in an equivalent state, they will always be able to perform the same actions.

It should be noted that in carrying out this proof, two types of reasoning were used. First, the laws of CIRCAL were used to develop a behavioural expression for IMP from its structural description and the behavioural specifications of its component parts. Then, laws of boolean logic were required to simplify predicates, to allow branches to be combined, and to show that the passed values and end states in the two behaviours IMP and BUF were equivalent. In general, a proof will involve reasoning about both the occurrence of actions and the equivalence of values or functions; these are referred to respectively as the temporal and functional aspects of a proof. It is only in the temporal aspect that CIRCAL provides a mechanism for carrying out the proof. In this example, the manipulations required to deal with the functional aspect were fairly straightforward; in more complicated example, the assistance of some sort of theorem prover might be desired. The use of such tools for verification is described in Section 5.2.6. However, dealing with the temporal aspect is both important and difficult, and it is apparent that CIRCAL is useful here. Whereas in this example the two aspects of the proof have been intermingled, ways in which they may be separated have been investigated, and are discussed in Section 5.2.4.

In this example, the implementation could be shown to be equivalent to specification in the fullest sense of the word, since the actions that can be performed by the constructed device are exactly those that can be performed by the specification. A more useful definition of equivalence in many circumstances is that of 'strong satisfaction'[Milne 85b]. Informally, this definition states that an implementation *satisfies* a specification if it can perform all the actions of the specification; it is also allowed to perform additional actions. This means that the implementation's behaviour need not be identical to the specification, as long as it includes all the behaviour of the specification. For deterministic devices (which include the vast majority of hardware), the key requirement for satisfaction is that

```
SPEC * IMP = SPEC
```

where '=' means 'can perform the same actions as'. The composition of SPEC with IMP ensures that any extra actions in IMP are removed before comparison with SPEC is attempted. Examples of the use of this approach to verification appear below.

5.2.2 Specification Techniques for Verification

It was shown in Section 5.1 that the development of specification languages and techniques in isolation can lead to unexpected problems when they are put to use. It is much more appropriate to develop a language by attempting to make use of it in some suitable application. One of the main applications of CIRCAL is for verification, and it is through attempting to use it for this purpose that many of the enhancements to the basic language and the techniques for using it have been developed and tested.

Fictitious Events

The effects of events which do not represent real changes of value (e.g an event {out<true} when the value on the port *out* is already true) have already been discussed in the context of simulation, and it was shown that descriptions of this type could dramatically reduce the efficiency of a simulation. In verification, efficiency may be less important, but this type of description can still cause problems, as the following example shows.

Example

A nand-gate can be specified in such a way that it generates an output event every time it receives an input event, as follows:

```
nandgate(x,y) <=
    {ina>p,out<nand(p,y)}nandgate(p,y)
  + {inb>p,out<nand(x,p)}nandgate(x,p)
  + {ina>p,inb>q,out<nand(p,q)}nandgate(p,q)
```

Explanation

The two state parameters represent the values on the two input ports *ina* and *inb* respectively. If a new value is input on either port, then the nand of this value and the value on the other port is output, regardless of whether it differs from the value currently on the port *out*.

This device is to be connected to the clock line of a positive-edge-triggered counter. Here it is vital that only true rising edges are accepted so that the counter really counts. Thus the counter would be specified as

```
counter(c,n) <= if not(c) then
   {clk<true,data<incr(n)}counter(true,incr(n))
 + if c then {clk<false}counter(false,n)
```

Explanation

The value on the clock port is represented by c and the counter's stored value by n. True rising edges can only occur if c is false. In this state, the event clk<true may occur, causing n to be incremented and placed on the output. A falling edge on the clock (clk<false) has no effect on the stored value.

The effect of composing the nand-gate, which can generate the fictitious events, with this counter which cannot accept them, might be expected to lead to problems. In fact, expansion of the expression that describes their interconnection leads to the following:

```
nandgate(x,y) [clk/out] * counter(c,n) <=
  {clk<true,data<incr(n),
    ina>p:and(not(c),eqs(nand(p,y),true))}
      nandgate(p,y)*counter(true,incr(n))
 + {clk<true,data<incr(n),
    inb>p:and(not(c),eqs(nand(x,p),true))}
      nandgate(x,p)*counter(true,incr(n))
 + {clk<true,data<incr(n),ina>p,
    inb>q:and(not(c),eqs(nand(p,q),true))}
      nandgate(p,q)*counter(true,incr(n))
 + {clk<false,
    ina>p:and(c,eqs(nand(p,y),false))}
      nandgate(p,y)*counter(false,n)
 + {clk<false,
    inb>p:and(c,eqs(nand(x,p),false))}
      nandgate(x,p)*counter(false,n)
 + {clk<false,ina>p,
    inb>q:and(c,eqs(nand(p,q),false))}
      nandgate(p,q)*counter(false,n)
```

Now consider what would happen if c were true. The only possible actions are the last three, since the others all contain input events where the parameter is restricted by a predicate which demands that c is false. The predicates in the last three actions all demand that any input events on the nand-gate be such that the new output of the nand-gate is false. Now suppose that the values on the ports *ina* and *inb* are both false. The output value of the nand-gate, and thus also c, will be true. An event on only one of *ina* or *inb* would not make the nand-gate's output become false. These events, therefore, cannot take place. If any device connected to the inputs of the gate attempted to perform those events, deadlock would result.

This example illustrates the danger of using an inconsistent approach to modelling. One device was specified, of necessity, in such a way that only genuine events (i.e. changes of value) could be accepted on its inputs. Another was specified, more conveniently, to accept any events and also to generate events that may not correspond to changes. The effect of wiring these two devices together was to introduce the possibility of deadlock in the model, even though the actual hardware that the two device descriptions are supposed to represent would not be able to deadlock. The solution is straightforward: the modelling philosophy that is adopted must be consistent; since there are situations such as the counter where non-genuine events are not acceptable, they must not be used in any device's specification.

A more general point to make here is that the language does not necessarily protect a user from writing incorrect specifications. It is to be hoped, however, that language features may be designed to reduce the opportunities for error. This was one of the main motivations for developing the language SuperC, described in Section 3.3.

Locked inputs

Some of the techniques presented in Section 3.2 were concerned with the prevention of 'locked' inputs. The justification given for this was simply that it contravenes the normal understanding of an input's behaviour in real hardware. In fact, these techniques arose from attempts to verify devices that were constructed from components whose inputs were occasionally locked. It had been hoped that, provided inputs were not required to change 'too quickly', then the specifications could be safely used. This turned out not to be the case, as the following example illustrates.

Example

A counter is to be constructed from cells, each of which controls one bit of the *n*-bit output and generates a carry signal for the next most significant cell. The cells can be specified constructively, with one part to control the output bit and another to generate the carry. Both parts are to produce an output change one time unit after an input change. The first part is described as follows:

```
output(c,d) <=
    if eqs(c,true) then
      {clk,t}{data<not(d),t}output(c,not(d))
  + if eqs(c,false) then {clk,t}output(c,d)
  + {cin>x:noteq(x,c),t}output(x,d)
  + {t}output(c,d)
```

Explanation

The parameters c and d represent the values on the carry input port *cin* and the output bit, *data*. For simplicity, the clock is assumed to be a pulse. The first branch indicates that if c is true then a clock event coinciding with a tick will cause the output value to change on the following tick. If c is false, a clock pulse has no effect. Other possible events are a new value being placed on *cin*, and an unaccompanied tick of the universal clock.

The carry generation part is described by the following expression:

```
cgen(c,d,z) <= if eqs(d,true) then
      {cin>x:noteq(x,c),t}{cout<x,t}
        cgen(x,d,x)
  + if eqs(d,false) then
      {cin>x:noteq(x,c),t}cgen(x,d,z)
  + if eqs(c,true) then
      {data>x:noteq(x,d),t}{cout<x,t}
        cgen(c,x,x)
  + if eqs(c,false) then
      {data>x:noteq(x,c),t}cgen(c,x,z)
  + {t}cgen(c,d,z)
```

Explanation

The values on the inputs *cin* and *data* are represented by the parameters c and d respectively, and the value on the carry output port *cout* by z. The most important aspect of the behaviour here is that changes on *cout* occur on the tick following the input changes that caused them, as indicated in the first and third branches, and that no other input events may take place until these output events have propagated.

Then the whole cell is described by

```
CELL <= cgen * output
```

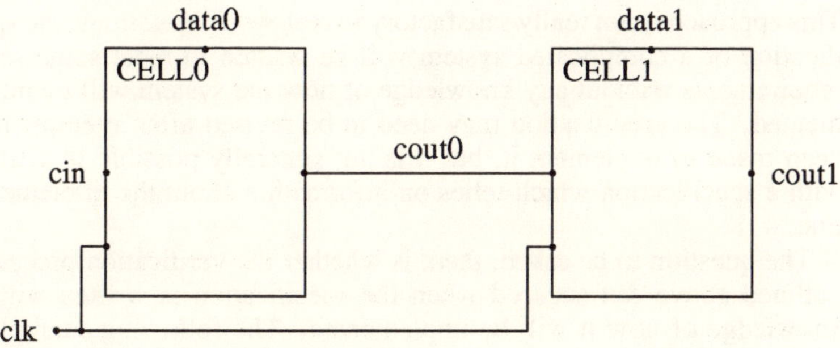

Figure 5.2: Two cells in an *n*-bit counter

and *n* of these cells must be connected together to construct an *n*-bit counter. Figure 5.2 shows the connection of two cells, with some re-labelling to enable correct wiring up of ports.

Now consider what happens if the values on the ports *cin*, *data0* and *cout0* are true when a clock event occurs. On the first tick after the clock pulse, the values on both *data0* and *data1* are complemented. This now means that the carry generation parts of both cells are committed to producing an event on their carry output ports, *cout0* and *cout1*, at the next tick. (This follows from the third branch of the description of cgen.) However, since the second cell is also connected to *cout0*, and is unable to accept an event on that port at the next tick, the system will reach a deadlock. Thus the conclusion to be drawn here is that the implementation is incorrect, since deadlock would certainly not be included in the counter's specification. In fact, if correctly described, this implementation would function correctly; therefore it is the specification technique which is at fault.

There are two ways in which this problem can be resolved. The obvious one is to use one of the techniques presented in Section 3.2 for the specification of devices with delays. Another approach would be to attach a *constraint* to the specification to ensure that the assumptions under which it was written (i.e. that inputs do not change 'too quickly') are actually satisfied. This approach is discussed in Chapter 7.

5.2.3   Verification Without Implementation Assumptions

In the verification of the two time-unit delay buffer at the beginning of Section 5.2.1, some assumptions were made about the implementation before writing the specification. Because of the simplicity of the example, the timing properties that would be exhibited by the implementation could be easily deduced, and the specification was tailored accordingly.

This approach is not really satisfactory in real-world situations; the specification of a complicated system will be written to meet some set of requirements without any knowledge of how the system will be implemented. The specification may need to be revised after attempts have been made to implement it, but it is not generally possible to start off with a specification which relies on information about the implementation.

The question to be asked, then, is whether the verification procedure outlined above can succeed when the specification is written without knowledge of how it will be implemented. The following example, a re-working of the buffer example of Section 5.2.1, demonstrates that it can.

Example

Rather than assuming that the delay across the buffer will be two ticks of the universal clock, another clock is introduced. The delay of the buffer will be one tick of this 'local' clock. The relationship between the local clock (whose ticks are designated by s) and the universal clock is not specified firmly; the only statement made about the two clocks is that the time between s ticks will be a non-zero multiple of the ticks of the universal clock. This relationship is described by the abstract device DEV as follows:

```
DEV  <=  {t}DEV + {s,t}DEV
```

This description ensures that ticks of the local clock, denoted by s, can only occur simultaneously with ticks of the universal clock, and that any number of t events (including zero) may take place between a pair of s events.

The buffer's specification is similar to that presented in Section 5.1.2:

```
BUF(x)  <=  {s}BUF(x)
   + {s,in>q:noteq(x,q)}BUF1(q)
BUF1(x)  <=  {s,out<x}BUF(x)
   + {s,in>q:noteq(x,q),out<x}BUF1(q)
```

Explanation

BUF(x) is the stable state, with x being the current value on the input. A tick of the local clock in this state produces no output change. State BUF1(x) is unstable, and so an s event will be accompanied by the placing of a new value on the output.

The proof now consists of trying to show that IMP1 satisfies SPEC, where those two behaviours are defined by

```
IMP1 <= (INV[mid/out] * INV[mid/in] - mid)
          * DEV
SPEC <= BUF * DEV
```

Note that both the buffer and the inverters must be connected to DEV to ensure that *t* and *s* are in the sorts of both the specification and the implementation.

The first part of IMP1 was calculated in Section 5.2.1 and given the name IMP. This is composed with DEV to give:

```
IMP1(m,n,d) <= IMP * DEV
= if eqs(not(n),d) then ({t}IMP1(m,m,not(n))
     + {t,in>p:noteq(p,m)}IMP1(p,m,not(n))
    + {s,t,in>p:noteq(p,m)}IMP1(p,m,not(n))
    + {s,t}IMP1(m,m,not(n)))
+ if noteq(not(n),d) then
   ({t,out<n}IMP1(m,m,not(n))
    + {t,in>p:noteq(p,m),out<n}
        IMP1(p,m,not(n))
   + {s,t,in>p:noteq(p,m),out<n}
        IMP1(p,m,not(n))
   + {s,t,out<n}IMP1(m,m,not(n)))
```

The next step in the proof is to show that IMP1 * SPEC is equal to SPEC. The expansion of SPEC gives

```
SPEC <= BUF * DEV
=  {t}(BUF(x) * DEV) + {s,t}(BUF(x) * DEV)
  + {s,t,in>q:noteq(x,q)}(BUF1(q) * DEV)
```

Therefore

```
IMP1 * SPEC <= if eqs(not(n),d) then
    ({t}(IMP1(m,m,not(n)) * BUF(x) * DEV)
    + {s,t}(IMP1(m,m,not(n)) * BUF(x) * DEV)
    + {s,t,in>p:noteq(p,m)}
        (IMP1(p,m,not(n)) * BUF1(p) * DEV))
```

which can be seen to be equal to SPEC, in terms of the actions that can be performed at this first level, provided eqs(not(n),d) is true (which is simply a requirement that the second inverter is initially in a stable state). Further expansions of the composition operator show that the implementation satisfies the specification fully.

The significant point here is that the specification was written without knowledge of the implementation and yet it was nevertheless possible to perform a successful verification of the implementation that was subsequently produced. The only assumption made about the behaviour of the implementation when writing the specification was that its delay would be an exact number of ticks of the universal clock. This is a reasonable assumption to make if the delays across all components in the implementation are to be measured in terms of those ticks. In more complicated examples, it may be important to be able to write a specification without guessing how the implementation will behave; this is illustrated in Chapter 6.

Two general principles emerge from this example. The first is that two separate but related clocks may be used to describe delays in such a way that the length of the delay is established by the implementation rather than by the specification. This technique should be applicable to a variety of situations, whenever a designer is not concerned with the exact length of time required by a piece of hardware to perform its function. The second principle, which is a generalisation of the first, is that it is possible to write a specification with an amount of uncertainty, this uncertainty being removed when the implementation is designed and verified. It is interesting to note that an implementation may contain more information than a specification and yet be shown to satisfy the specification. This is an encouraging result, as it demonstrates the feasibility of showing the equivalence of circuit descriptions that contain different amounts of information.

5.2.4 *Separating Timing and Function*

In the example of Section 5.2.1 two aspects of the proof process, temporal and functional, were identified. CIRCAL does not provide any direct means for carrying out the functional aspects of the proof. However, it would be useful to be able to split the verification problem into its two parts, using suitable reasoning frameworks to deal with each. An attempt to formalise this separation procedure is presented in this Section.

Whether it is required to prove equivalence or satisfaction between specification and implementation, it is ultimately necessary to show that two behavioural expressions are equivalent. In most circumstances, each behaviour will be a deterministic sum of branches. It may be possible to show at once that they are not equivalent, if there is a guard in one behaviour that cannot be matched with any guard in the other. In the introductory example, branches were combined until the guards in the two behaviours were identical. However, it is possible for non-identical guards to match. In general, two guards can be matched if they contain events on exactly the same ports and the events match. Any type of output event (value, parameter or function) matches any other output event (on the same port); two input events match if there is some set of values which satisfies both the attached predicates; and two synchronisation events match.

Using these rules, several guards in one behaviour may match a single guard in the other (as was the case in Section 5.2.1 before the branches of IMP were combined). In order to show that a number of branches in one behaviour are equivalent to a single branch in the other, certain relationships between the predicates and functions in the two behaviours must be satisfied. The derivation of these relationships in the case where the guard consists of a single output event is presented below.

The two behaviours are

```
A(x)  <= if f1(x) then {p<g1(x)}A(h1(x))
     + other branches
B(y)  <= if f2(y) then {p<g2(y)}B(h2(y))
     + if f3(y) then {p<g3(y)}B(h3(y))
     + other branches
```

In each case, the 'other branches' contain guards involving ports other than p. All the fs are predicates; the gs and hs are functions. Although each behaviour is parameterised over only one variable, the following argument easily extends to a number of parameters.

Assuming that A describes the specification and B the implementation, the first requirement is that, for any value of x, there is some value of y such that the value passed on p will be the same for both behaviours. That is,

$$\forall x \quad (\exists y. \quad (f1(x) \wedge f2(y) \supset g1(x) = g2(y))$$
$$\wedge (f1(x) \wedge f3(y) \supset g1(x) = g3(y))$$
$$\wedge (f1(x) \equiv f2(y) \vee f3(y)))$$

The third line ensures that there will be a valid branch in the behaviour B whenever there is a valid branch in A.

The above formulae, if satisfied, ensure that the initial actions of A and B are identical. To ensure that subsequent actions are identical, some relations need to be established between the end states of the two behaviours. In short, the relationships defined above that must exist between x and y must also exist between the parameters of the end states. If the three lines following $\exists y$ above are represented by a single predicate P(x,y), then the following relationship must also be satisfied:

$$(f1(x) \wedge f2(y) \supset \quad P(h1(x), h2(y)))$$
$$\wedge \quad (f1(x) \wedge f3(y) \supset \quad P(h1(x), h3(y)))$$

The first line ensures that the second branch of B leads to states in which the two behaviours can perform identical actions; the second line takes care of the second branch of B.

This result demonstrates that it is possible to show that two behavioural expressions are equivalent by proving a logical formula that contains no reference to time or the ordering of events. That is, the proof problem has been effectively divided into the temporal part, which is handled by CIRCAL, and the functional part, which may be proved either manually or with some mechanical assistance. The above example is far from being completely general, but the basic principles of matching guards and establishing that passed values are equivalent in various circumstances could be readily extended to other situations. Since the verification of timing properties is often a major constituent of a correctness proof, the ability of CIRCAL to deal with them may make a significant contribution to verification.

5.2.5 *Verification vs. Exhaustive Simulation*

Only exhaustive simulation, in which a device is simulated under all possible combinations of input patterns and internal states, can provide the same certainty of an implementation's correctness that can be achieved by formal verification. It is widely accepted that exhaustive simulation is not an acceptable validation technique, as it is prohibitively time-consuming for devices of realistic complexity. Having seen how much effort is required to perform a very simple verification exercise, however, one might be tempted to draw the same conclusion about verification. Fortunately, this turns out not to be the case.

The development of the CIRCAL language as described in Chapter 3 has meant that specifications in the language need not grow exponentially in size with the number of states of a device. This has been a necessary first step in reducing the complexity of the verification task. As was discussed above, the proof techniques that are employed for devices specified in CIRCAL deal quite well with timing and sequencing aspects of behaviour; the remaining complexity is in the area of functional correctness. There are, however, many situations in which functional correctness can be established with less effort than would be required if exhaustive simulation was performed. The equivalent of exhaustive simulation in theorem proving is the technique of case analysis; there are, fortunately, many other proof techniques, such as mathematical induction, that may be used in certain circumstances and are far more efficient.

One main advantage of mathematical proof for validation is that once a particular theorem has been proved for one part of a system, it can be re-used any number of times in the rest of the system. Furthermore, many hardware devices have a regular structure that may be parameterised by the number of bits in the data words that are processed. In such a case, a single verification of a general *n*-bit device will suffice for any number of specific instances of the device in the design of a whole chip. Such savings of effort are not generally possible when validation is performed by simulation.

Some other verification efforts have concentrated on the problems of tackling complexity. A selection of these projects is presented and discussed below.

5.2.6 *Other Approaches*

Whereas CIRCAL is a framework created specifically for the purposes of hardware description and verification, a number of other research efforts have attempted to use existing frameworks to reason about hardware. One such framework in which verification has been attempted with considerable success is that of the Boyer-Moore logic [Boyer 81] (a form of first-order logic), notably by Hunt [Hunt 86a, Hunt 86b]. A mechanised-theorem prover was already available for this logic and Hunt was able to find a way of applying it to the description and verification of hardware, the largest example to be tackled being a 16-bit microprocessor of quite considerable complexity. Some simplifying assumptions were necessary to achieve this result, in particular a fairly simplified model of timing and the use of bistate logic throughout the design. The bottom level of hierarchy in this design was that of logic gates. This verification represented quite a landmark in terms of the complexity of the device verified.

Another noteworthy framework in which hardware verification is being investigated is higher-order logic. The use of this logic for the description of hardware has already been examined in Section 2.2.4. Hanna [Hanna 85] was one of the first to use this approach, representing hardware devices as predicates over waveform specifications. The treatment of time in this work is very rigorous, and the verification that is performed relies on a minimum of assumptions about the implementation; in this way it is similar to the approach proposed in Section 5.2.3. The penalty of this rigorous approach, however, is that the complexity of the verification task becomes quite high. Consequently, the examples that have been tackled in this way so far have been at a fairly low level of abstraction.

Higher-order logic has also been applied to verification using the HOL system [Gordon 85]. The availability of a sophisticated, mechanised proof system to support this descriptive framework (described in Section 2.2.4) has led to much interest in it. A notable achievement in the use of this system was the verification, between two fairly high levels in the design hierarchy, of the VIPER microprocessor [Cohn 87], a real machine that has been fabricated. A considerable amount of research has concentrated on dealing with temporal aspects of proof. In HOL, for example, Melham has worked on temporal abstraction to reason about hardware behaviour on different time scales [Melham 87], an approach which bears some similarity to that described in Section 5.2.3; Herbert has described work using temporal constraints to separate temporal and functional aspects of proof [Herbert 88], a problem that was also addressed in Section 5.2.4. In general the treatment of timing is not quite as detailed as that proposed by Hanna and Daeche, thereby allowing the treatment of more complex devices.

Temporal logic, described in Section 2.2.5, is also used for formal verification. A mechanised theorem-prover for this logic has been developed and used for this purpose [Moszkowski 85]. Since the language has special operators for temporal aspects of behaviour, the separation of timing and function is rather straightforward.

The above research efforts represent only a fairly small cross-section of those being undertaken; an excellent and comprehensive survey of the subject is found in [Camurati 87]. The success with which these attempts have met illustrates the fact that complexity need not be an insurmountable obstacle to verification, while the diversity of areas in which they have been applied demonstrates the extent to which the nature of languages influences the verification task. Furthermore, the failure of any formal verification method to replace simulation as the predominant validation technique used by designers emphasises the need for continued research into formal techniques.

5.2.7 *Verification of Transformations*

In Section 4.3.2 the issue of validating automated designs was discussed. While it is perfectly possible to validate the output of the automation tool just as if it had been built by a human designer, it is preferable to validate the tool itself; in this way all devices produced by the tool may be guaranteed correct, at a cost of only a single validation effort.

The formal verification of a very simple design automation tool is the subject of a paper by Milne [Milne 83b]. The tool is described formally as a transformation which maps logical expressions to layout information. A simple language NE is used to represent the logical expressions, and another language LL to represent layout. The transformation is represented by a function L. Functions to map NE expressions and LL expressions to CIRCAL are defined as N and S respectively. The task of validating the transformation is therefore to establish that

$$\forall n \in \text{NE}. S \circ L\,(n) = N\,(n)$$

Because of the simplicity of the two languages used, this task is fairly straightforward. It remains to be seen whether it could be applied to realistic silicon compilers or other tools. However, it certainly establishes the value of using languages with well-understood semantics as the input and output languages of DA tools, as it is only with such languages that this type of validation could be attempted.

5.3   Summary

In this Chapter the validation task, consisting of the composition of implementation behaviours and their comparison with specifications in an attempt to establish that a design step has been correctly performed, has been presented. Two approaches to the task were introduced: simulation and formal, mathematical verification. The former approach is popular because it places fewer requirements on the specification language to be used, as well as simply being better established as a validation technique. Verification, however, offers the significant advantage over simulation of guaranteed circuit correctness.

Since simulation consists of establishing the response of a circuit to selected input stimuli, some way of specifying these stimuli is required. A case was made for using a standard behavioural hardware description language to do this. Two approaches to the simulation of constructed devices were discussed. The more common of these involves establishing the possible responses of each component at each simulation step and using this information to calculate the resultant action. The only requirement on the language in this case is that the responses of behaviourally specified components to input stimuli can be established. This requirement is satisfied by the majority of behavioural hardware description languages.

The fact that CIRCAL enables behaviours to be constructed mathematically, together with its event-based model of behaviour, make it suitable for a different approach to simulation. Expansion of CIRCAL's composition operator leads to behavioural expressions from which the responses of constructed circuits to input stimuli can readily be calculated.

The ability to construct behaviours mathematically is essential to formal verification. Some small examples were used to show how proofs could be carried out using CIRCAL. Two aspects of verification, temporal and functional, were identified, and it was seen that CIRCAL is able to provide real assistance in only the first of these. However, it also assists in separating the two aspects in a way that would facilitate the solution of the functional aspect by some other means. It was shown that a specification with some uncertainty could be matched against an implementation in which the uncertainty was removed. This means that specifications can be written using a minimum of assumptions or prior knowledge about the implementation (as they would normally be in real design situations) without inhibiting the verification task.

One of the main concerns in the development of verification techniques is the complexity that is encountered. While this is a problem, it was seen to be less severe than the computational complexity of exhaustive simulation which prohibits its use in realistic situations. The development of languages for verification was seen as an essential first step in combating complexity. Some of the other approaches to verification that have dealt with the complexity problem to some extent were discussed.

The attempts to carry out both simulation and verification using descriptions written in CIRCAL illustrated the need to test and develop specification techniques by putting specifications to work. Some of the specification techniques of Chapter 3 that seemed intuitively correct were justified by examples that demonstrated the dangers of ignoring those techniques. In particular, it was seen that a verification of a quite reasonable implementation could fail if the device specifications were not sufficiently accurate. The general principle that emerges here is that the modelling philosophy that is adopted must be consistent and as realistic as possible.

SIX

EXAMPLE: A SIMPLE COMPUTER

In the preceding three Chapters, the three main tasks of the proposed
methodology (specification, design and validation) have been presented
and discussed. The examples used to illustrate the points in these chapters have been necessarily quite simple. Furthermore, the examples have
generally been chosen to illustrate a single point about just one of the
tasks. It is the aim of this Chapter to illustrate the way in which these
tasks fit together in the execution of a single design step by tackling a
significantly larger example than those presented previously.

The system that has been chosen for examination is based on a simple computer that was originally verified using the LCF_LSM system
[Gordon 81a]. The same implementation has subsequently been verified
using HOL [Joyce 86]. In this Chapter, a specification of the computer
using enhanced CIRCAL will be developed. The verification of the computer as specified would be extremely involved and rather too lengthy to
present here. Therefore a simplified version of the problem will be posed.
A way of approaching the design that assists subsequent verification will
be presented, followed by the verification itself. Before commencing
formal treatment of the problem, however, it will be helpful to examine
an informal statement of it.

6.1 Informal Specification

The computer has a 13-bit program counter, a 16-bit accumulator and
an $8k \times 16$-bit random access memory (RAM). There is a 16-bit input
port for the loading of values into the registers, and a 4-position switch
to determine the mode of operation of the computer. There is also a button, the use of which is dependent on the current mode and is described
below. The values stored in each register may be observed at a pair of
output ports, one 13 bits wide and the other 16. A single bit output port
idle is also provided; its function is also described below. A 'black-box'
diagram of the computer appears in Figure 6.1. Ports are illustrated in
various widths to indicate the width of word (in bits) that can be passed

along it. The ports on the left are inputs, those on the right are outputs.

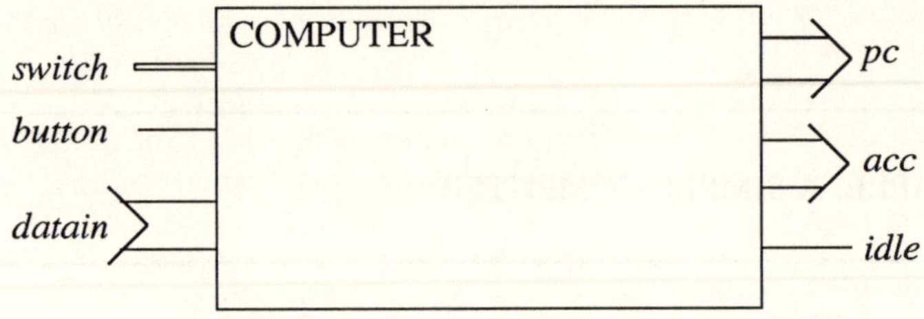

Figure 6.1: Black-box diagram of computer

The 4-position switch may be considered as an input port of type int2, this being a type containing all the integers that may be represented by 2 bits. The value n on this port when the button is pushed, provided *idle* is true, determines the operation of the computer as follows:

$n = 0$: The program counter register is loaded with the value obtained by truncating the value on the 16-bit input port *datain* to 13 bits.

$n = 1$: The accumulator is loaded with the value on the input.

$n = 2$: The contents of the accumulator are loaded into the memory location whose address is in the program counter.

$n = 3$: The program stored in the memory is executed, beginning at the address that is in the program counter.

Execution of the program is the most involved part of the computer's operation and is described below.

Opcode	Mnemonic	Meaning
000	HALT	Stop execution of program
001	JMP L	Jump to L
010	JZRO L	Jump to L if acc. contains 0
011	ADD L	Add contents of L to acc.
100	SUB L	Subtract contents of L from acc.
101	LD L	Load contents of L into acc.
110	ST L	Store contents of acc. at L.
111	SKIP	Skip to next instruction

Table 6.1: Meanings of the opcodes

A 16-bit word is interpreted as an instruction by separating it into an opcode (the most significant 3 bits) and a 13 bit operand. The meanings of the 8 opcodes are explained in Table 6.1, with L being the operand, interpreted as an address. The value on *idle* becomes false when execution begins and true when execution stops. Execution may be stopped either by a HALT instruction or by pushing the button, in which case execution of the current instruction will be completed before stopping.

That describes fully the operation of the computer. The following Section shows how this description can be captured as a formal specification in enhanced CIRCAL.

6.2 *Formal Specification*

Before writing the actual CIRCAL descriptions for the computer, some type and function definitions are necessary. The required types are as follows:

- `int2`, `int13` and `int16`. These are defined as 2-, 13- and 16-bit integers respectively.
- `memory`. This type represents the 8k × 16-bit random access memory.
- `opcode`. A variable of this type may be any one of the 8 opcodes.

Functions required for the manipulation of objects of the above types are:

- `truncate:  int16 -> int13`. Truncate a 16-bit integer to 13 bits.
- `store:  memory * int16 * int13 -> memory`. Return a new memory state, obtained by storing a 16-bit integer at a 13-bit address.
- `fetch:  memory * int13 -> int16`. Obtain the value stored at an address in a memory.
- `getop:  memory * int13 -> opcode`. Extract the opcode from a 16-bit integer stored at a location in memory.
- `getadd:  memory * int13 -> int13`. Extract the address from the 16-bit integer stored at a location in memory.
- `getarg:  memory * int13 -> int16`. Extract the low 13 bits from a location in memory and fetch the 16-bit integer stored at the resultant address.
- `incr:  int13 -> int13`. Increment a 13-bit integer.
- `add16:  int16 * int16 -> int16`. Add two 16-bit integers.
- `sub16:  int16 * int16 -> int16`. Subtract a 16-bit integer from a 16-bit integer.

Formal definitions of these types and functions may be found in Appendix D.

A formal specification of the computer may now be formulated in enhanced CIRCAL. In order to simplify this process, it is helpful to make some assumptions about the way in which the device will be used. The informal specification does not, for example, state what happens if the switch is moved to a new position and the button is pushed simultaneously. Therefore, it is not particularly helpful to include in the formal specification a statement of what would happen in this instance. While it was seen in Section 5.2.2 that the writing of specifications that disallow certain input events could cause problems in verification, it will be demonstrated in the following Chapter that the placing of suitable restrictions on a device's environment can overcome these problems. For the moment, the implicit assumption is that simultaneous input events will not occur. In the next Chapter the formalisation of such assumptions will be presented.

Because it is rising edges on the port *button* that are significant, the edge detector idea introduced in Section 3.2 can be used to reduce the number of states and the interleaving of events in the main part of the specification. The edge detector box could be described as follows:

```
EDET  <=  {t}EDET + {button<true}EDET1
        + {t,button<true, but}EDET2
EDET1 <=  {t,but}EDET2 + {button<false}EDET
        + {t,button<false}EDET
EDET2 <=  {t}EDET2 + {button<false}EDET
        + {t,button<false}EDET
```

Explanation

This is a little more complicated than the box of Section 3.2, as it ensures that but events occur synchronously with ticks, even if rising edges on the *button* port do not. In the state EDET the button is yet to be pushed, i.e. the value on *button* is false. Ticks may be passively accepted, or if the button is pushed between ticks then the device moves into state EDET1. If the button is pushed at the same time as a tick, then a pulse is generated immediately on *but*. In state EDET1 the next tick causes the pulse on *but* to be produced, unless the button is released first, in which case the device simply returns to the initial state EDET. In the third state, EDET2, the device simply waits for the release of the button before returning to EDET.

If this aspect of the computer's behaviour were included in the main part of the description, there would be quite a large number of ways in which the various events on *button* could interleave with events on other ports. This would complicate the description of the whole computer significantly. Separating this part of the behaviour from the main description considerably reduces the effort of writing that description.

The main specification can now be formulated. The states denoted by `computer` are the 'idle' states, in which the value on *idle* is true and pushing the button will cause an action dependent on the position of the mode switch. The state `computer1` represents the execution of instructions in memory; in this state, *idle* is false and pushing the button will halt execution. The state parameters `p`, `ac` and `dat` are used to represent the values on the ports *pc* (program counter), *acc* (the accumulator) and *datain* respectively, and an additional parameter of type `memory` represents the state of the RAM. The last parameter represents the position of the mode switch.

```
computer(mem,p,ac,dat,0) <=
   {but,pc<truncate(dat),t}
      computer(mem,truncate(dat),ac,dat,0)
 + {datain>x:noteq(x,dat),t}
      computer(mem,p,ac,x,0)
 + {mode>x:noteq(x,0),t}
      computer(mem,p,ac,dat,x)
 + {t} computer(mem,p,ac,dat,3)

computer(mem,p,ac,dat,1) <=
   {but,acc<dat,t}computer(mem,p,dat,dat,1)
 + {datain>x:noteq(x,dat),t}
      computer(mem,p,ac,x,1)
 + {mode>x:noteq(x,1),t}
      computer(mem,p,ac,dat,x)
 + {t}computer(mem,p,ac,dat,3)

computer(mem,p,ac,dat,2) <=
   {but,t}computer(store(mem,ac,p),p,ac,dat,2)
 + {datain>x:noteq(x,dat),t}
      computer(mem,p,ac,x,2)
 + {mode>x:noteq(x,2),t}
      computer(mem,p,ac,dat,x)
 + {t}computer(mem,p,ac,dat,3)

computer(mem,p,ac,dat,3) <=
   {but,t,idle<false}
      computer1(mem,pc,dat,dat,3)
 + {datain>x:noteq(x,dat),t}
      computer(mem,p,ac,x,3)
 + {mode>x:noteq(x,3),t}
```

```
         computer(mem,p,ac,dat,x)
  + {t}computer(mem,p,ac,dat,3)

computer1(mem,p,ac,dat,m) <=
    if eqs(getop(mem,p),HALT) then
    {t,idle<true}
        computer(mem,p,ac,dat,m)
  + if eqs(getop(mem,p),JMP) then
    {t,pc<getadd(mem,p)}
        computer1(mem,getadd(mem,p),ac,dat,m)
  + if and(eqs(getop(mem,p),JZRO),eqs(ac,0))
      then {t,pc<getadd(mem,p)}
        computer1(mem,getadd(mem,p),ac,dat,m)
  + if and(eqs(getop(mem,p),JZRO),noteq(ac,0))
      then {t,pc<incr(p)}
        computer1(mem,incr(p),ac,dat,m)
  + if eqs(getop(mem,p),ADD) then
      {t,pc<incr(p),
                acc<add16(getarg(mem,p),ac)}
        computer1(mem,incr(p),
                add16(getarg(mem,p),ac),dat,m)
  + if eqs(getop(mem,p),SUB) then
      {t,pc<incr(p),
                acc<sub16(ac,getarg(mem,p))}
        computer1(mem,incr(p),
                sub16(ac,getarg(mem,p)),dat,m)
  + if eqs(getop(mem,p),LD) then
      {t,pc<incr(p),acc<getarg(mem,p)}
        computer1(mem,incr(p),
                getarg(mem,p),dat,m)
  + if eqs(getop(mem,p),ST) then {t,pc<incr(p)}
        computer1(store(mem,ac,p),incr(p),
                ac,dat,m)
  + if eqs(getop(mem,p),SKIP) then
      {t,pc<incr(p)}
        computer1(mem,incr(p),ac,dat,m)
  + {but,t,idle<true}computer(mem,p,ac,dat,m)
```

An interesting point to note here is that the time taken for the execution of each instruction is assumed to be the interval between two ticks of the universal clock. The implementation may be expected to take varying amounts of time to execute different instructions, so there will need to be some way of relating the two different time scales.

6.2.1   A Reduced Problem

The above specification illustrates how the descriptive power of enhanced
CIRCAL is sufficient to cope with the quite complicated example of a
small computer. In order to illustrate the design and especially the verification tasks in the space available here, however, it is necessary to reduce
the complexity of the example. One simplification is to concentrate on
only the execution of stored programs, ignoring the three modes of operation in which data is input to registers or memory using *datain*. A further
simplification would be to reduce the instruction set to 4 instructions,
say LD, ADD, SUB and JZRO. (This could be just enough to execute
some useful programs stored in Read-Only Memory (ROM).) Since the
opcode field needs only 2 bits now, addresses could be increased to 14
bits, and the ML functions and datatypes described above could be modified accordingly. The state of memory, mem, cannot now be modified,
so need not be included in the list of parameters for the computer. When
it appears as an argument to a function, it may now be considered as a
constant. As the only mode of operation of the computer is execution
of programs, there is no need for a button to halt it. The new computer
therefore has sort {*pc, acc*} and can be specified as follows:

```
newcomputer(p,ac)  <=
    if eqs(getop(mem,p),ADD) then
      {t,acc<add16(ac,getarg(mem,p)),
                    pc<incr(p)}
        newcomputer(incr(p),
                    add16(ac,getarg(mem,p)))
   + if eqs(getop(mem,p),SUB) then
       {t,acc<sub16(ac,getarg(mem,p)),
                    pc<incr(p)}
         newcomputer(incr(p),
                    sub16(ac,getarg(mem,p)))
   + if eqs(getop(mem,p),LD) then
       {t,acc<getarg(mem,p),pc<incr(p)}
         newcomputer(incr(p),getarg(mem,p))
   + if and(eqs(getop(mem,p),JZRO),eqs(ac,0))
       then {t,pc<getadd(mem,p)}
           newcomputer(getadd(mem,p),0)
   + if and(eqs(getop(mem,p),JZRO),noteq(ac,0))
       then {t,pc<incr(p)}
           newcomputer(incr(p),ac)
```

Explanation

This specification is very similar to that for the execution mode of the full computer. In each branch, the current instruction is obtained using `getop` and compared with one of the possible opcodes. In the case of JZRO, the action that results also depends on the value in the accumulator, represented by the parameter `ac`. Following the conditional in each branch is a guard consisting of a tick, an event on *pc* (either incrementing its value or loading it with a new value from memory), and possibly an event on *acc*. The design and verification of the computer specified in this way will be described in the following Sections.

As an aside, it should be noted that in this specification, as in the full specification of Section 6.2, there is the potential to perform fictitious output events, for example if `getarg(mem,p)` is 0 during an ADD operation. The consequences of this omission, which has been made to simplify the presentation of the example, are made clear below.

6.3 Design

In the verification of the computer using both LCF_LSM [Gordon 81a] and HOL [Joyce 86] the design step was a large one. That is to say, the amount of structural information that was added at the lower level of the design hierarchy was relatively large. This is apparent simply from consideration of the large number and relatively small size of components into which the black-box computer was partitioned in each of these exercises. The large gap between the specification level and the implementation level makes the verification task significantly more complicated. As proposed in Section 4.1.1, if design is to assist verification, then smaller design steps should be taken. The contrast between these two approaches is illustrated in Figure 6.2.

In this Section, the design of the simplified example, `newcomputer`, will be presented. This will be done using a small design step, the black box being partitioned into just two fairly large boxes. Even though the example being tackled here is simpler than the computer first described, the boxes with which it is to be implemented are still larger than those of the implementation level of Gordon's design. In the following Section, the way in which this assists the verification task will be demonstrated.

The first phase of design is partitioning. The simplest partitioning that can be made is to split the computer into two boxes, a control part and a data part. The latter handles the transfer of data between memory and registers, including the fetching of instructions, while the former provides control signals for the data part depending on the current instruction and possibly the accumulator value. This suggests that there should be a port to carry control signals from `cpart` to `dpart`, and ports to convey the values of the current instruction and the accumulator from

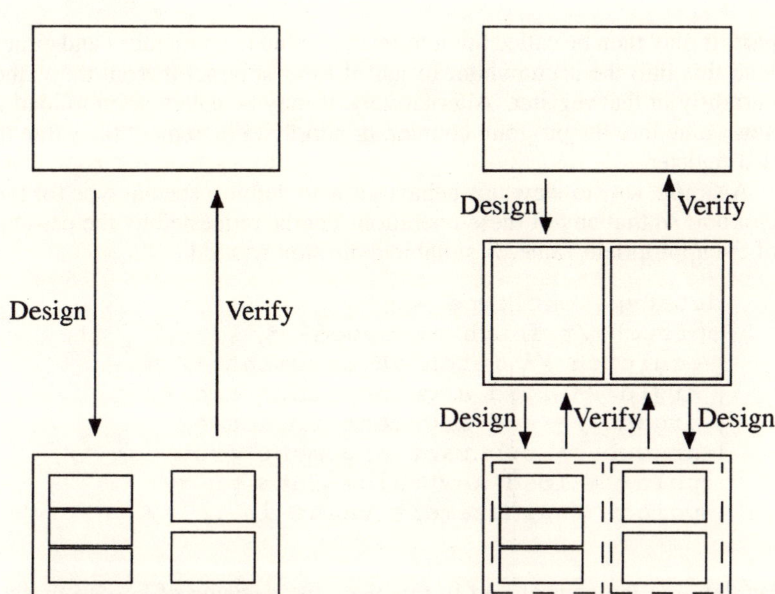

Figure 6.2: Different design step sizes

`dpart` to `cpart`. This structural information is conveyed by the following CIRCAL expression and is illustrated in Figure 6.3.

```
computerimp <= cpart * dpart - cntl - cur
```

6.3.1   The Data Part

The partitioning phase out of the way, the next step is to describe the boxes of the implementation. In order to define the behaviour of `dpart` it is helpful to consider what functions it must perform. At the start of each instruction cycle it must fetch an opcode and feed this to the control

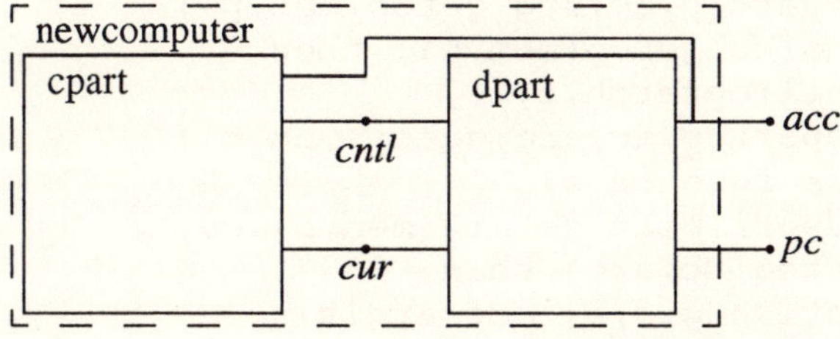

Figure 6.3: Structural view of newcomputer implementation

part. It may then be called upon to fetch a value from memory and either load this into the accumulator or add it to or subtract it from the value currently in that register. Alternatively, it may be called upon to load a new value into the program counter, or simply to increment the value in that register.

A simple way to view this behaviour is to define a special type for the *cntl* port so that any of these operations can be requested by the passing of the appropriate value. A suitable definition would be

```
datatype cntltype =
opfetch /* fetch an opcode */
| argfetch /* fetch an argument */
| argld /* load argument into acc */
| argadd /* add argument to acc */
| argsub /* subtract argument from acc */
| pcld /* load new value into pc */
| pcincr /* increment value in pc */
```

By defining the control port in this way, the decision of how to implement it physically (e.g. with 7 wires of which only one may be true at one time) is left until later, when the internal structure of dpart is defined. This is an example of the way in which finer design steps assist design by allowing the postponement of design decisions until more information becomes available.

It is now quite straightforward to define the behaviour of dpart. In order to enable its timing characteristics to be reconciled with those of the specification, the technique of Section 5.2.3 is adopted, with delays at this level of abstraction being related to a different clock whose ticks are denoted by s.

```
dpart(p,ac,arg,cu) <=
    if noteq(cu,getop(mem,p)) then
      {cntl<opfetch,s}{cur<getop(mem,p),s}
        dpart(p,ac,arg,getop(mem,p))
  + if eqs(cu,getop(mem,p)) then
      {cntl<opfetch,s}{s}
        dpart(p,ac,arg,getop(mem,p))
  + {cntl<argfetch,s}
        dpart(p,ac,getarg(mem,p),cu)
  + if noteq(ac,arg) then
      {cntl<argld,s}{acc<arg,s}
        dpart(p,arg,arg,cu)
  + if eqs(ac,arg) then
```

```
    {cntl<argld,s}{s}dpart(p,arg,arg,cu)
 + if noteq(arg,0) then
      ({cntl<argadd,s}{acc<add16(acc,arg),s}
          dpart(p,add16(acc,arg),arg,cu)
      + {cntl<argsub,s}{acc<sub16(acc,arg),s}
          dpart(p,sub16(acc,arg),arg,cu))
 + if eqs(arg,0) then
      ({cntl<argadd,s}{s}dpart(p,acc,arg,cu)
      + {cntl<argsub,s}{s}dpart(p,acc,arg,cu))
 + if noteq(p,getadd(mem,p)) then
      {cntl<pcld,s}{pc<getadd(mem,p),s}
          dpart(getadd(mem,p),ac,arg,cu)
 + if eqs(p,getadd(mem,p)) then
      {cntl<pcld,s}{s}dpart(p,ac,arg,cu)
 + {cntl<pcincr,s}{pc<incr(p),s}
          dpart(incr(p),ac,arg,cu)
 + {s}dpart(p,ac,arg,cu)
```

Explanation

The parameters p and ac represent the program counter and accumulator as before, while arg represents the fetched argument following an argfetch signal, and cu is the value on the current instruction port *cur*. In each branch except the third and last the arrival of a new control signal on one tick is followed by an output event on the next, unless the current value on the output port equals the new value that would be placed there. The third branch involves no output event but simply a change in the internal state parameter arg and takes only one tick of the clock, while the last branch simply states that the device will remain in a steady state as time passes if no *cntl* input is provided.

There are a few things to note in this description. One is that the rule of avoiding locked inputs has been ignored, since an event on the *cntl* input is usually followed by some output event on the next tick. The justification for this is that the only device that will ever attempt to communicate on the *cntl* port is the control part, so as long as it is designed with this limitation in mind, there should be no problem. In fact, this is an example of a situation in which a constraint is being applied to a device's environment — the formal treatment of such situations is discussed in the next Chapter. An additional assumption made about the behaviour of cpart is that it will not generate fictitious events on the *cntl* port, which again should be borne in mind when writing its specification.

6.3.2 *The Control Part*

The data part implements the various functions of the computer; it is the
role of the control part to ensure that these functions occur at the appropriate time. That is, given inputs of the current instruction and accumulator
value, it must generate the correct sequence of values on the control port.
This may be specified formally as follows:

```
cpart(cu,ac) <= {cntl<opfetch,s}cpart1(cu,ac)
cpart1(cu,ac) <=
    {cur>x:noteq(x,cu),s}cpart2(x,ac)
 + {s}cpart2(cu,ac)
cpart2(cu,ac) <=
   if or(eqs(cu,ADD),
             or(eqs(cu,SUB),eqs(cu,LD))) then
     {cntl<argfetch,s}cpart3(cu,ac)
 + if and(eqs(cu,JZRO),eqs(ac,0)) then
      {cntl<pcld,s}cpart6(JZRO,0)
 + if and(eqs(cu,JZRO),noteq(ac,0)) then
      {cntl<pcincr,s}cpart6(JZRO,ac)
cpart3(cu,ac) <=
   if eqs(cu,ADD) then
      {cntl<argadd,s}cpart4(cu,ac)
  + if eqs(cu,SUB) then
      {cntl<argsub,s}cpart4(cu,ac)
   + if eqs(cu,LD) then
      {cntl<argld,s}cpart4(cu,ac)
cpart4(cu,ac) <= {s}cpart5(cu,ac)
   + {s,acc>x:noteq(x,ac)}cpart5(cu,x)
cpart5(cu,ac) <= {cntl<pcincr,s}cpart6(cu,ac)
cpart6(cu,ac) <= {s}cpart(cu,ac)
```

Explanation

In the initial state `cpart`, the control signal is given to fetch the next
opcode. In `cpart1`, the opcode is read in, the second term of the choice
sum allowing for the possibility that the new opcode will be the same as
the last one. `cpart2` is the instruction decode stage. For all instructions
other than the jump, the first step is to fetch the argument from memory,
so the necessary control signal is `argfetch`. If the accumulator contains 0, then the jump requires a new value to be loaded into *pc*; otherwise
pc must simply be incremented. In either case, the next state is `cpart6`
and, after a delay of one tick, the instruction is now complete and the
control part returns to its initial state. For the other 3 instructions, a further decoding is required, this being done in state `cpart3`. The control

signal to add, subtract or load the argument into the accumulator is generated; then in state `cpart4` there is another delay of one tick. Since the operation may cause a new value to be placed into the accumulator, this possibility is also accounted for in this state. In state `cpart5` the program counter is incremented and after one further tick the device returns to its initial state.

Having partitioned and described the implementation completely, it now needs to be shown that its behaviour satisfies that of the specification. This task is described in the following Section.

6.4 Verification

As in Section 5.2.3, the timing characteristics of the implementation are measured with respect to a different clock from those of the specification. Therefore, the first step of the verification proof is to define the relationship between the two clocks. The requirement is that the implementation will, after a certain number of s ticks, perform the correct action (as laid down by the specification) at the same time as a t tick. As before, it is assumed that there may be one or more s ticks between the t ticks. At the outset of the verification, it is not known exactly how many s ticks there will be, so the device that defines the relationship between the two clocks can be described in such a way as to allow this uncertainty:

```
DEV(n) <= if noteq(n,1) then {s}DEV(decr(n))
          + if eqs(n,1) then {s,t}DEV(M)
```

The initial value of n is left unspecified, as is the value of M to which the tick-counting parameter is assigned. This uncertainty will be removed as the verification proceeds.

In order for the implementation to satisfy the specification, it is required that the actions which may be performed by

```
IMP <= (cpart * dpart - cntl - cur) * DEV
          * newcomputer
```

are the same as the actions of

```
SPEC <= DEV * newcomputer
```

The behaviour of IMP is obtained by expansion of the composition and abstraction operators. The first step is to expand `cpart * dpart` using the synchronisation rules of Section 3.1.6:

```
cpart(cu,ac) * dpart(p,ac,arg,cu) =
    if noteq(cu,getop(mem,p)) then
      {cntl<opfetch,s}(cpart1(cu,ac) *
```

```
          {cur<getop(mem,p),s}
              dpart(p,ac,arg,getop(mem,p)))
      + if eqs(cu,getop(mem,p)) then
          {cntl<opfetch,s}(cpart1(cu,ac) *
          {s}dpart(p,ac,arg,getop(mem,p)))
```

Inserting the definition of `cpart1` and expanding again gives:

```
cpart(cu,ac) * dpart(p,ac,arg,cu) =
    if noteq(cu,getop(mem,p)) then
      {cntl<opfetch,s}{cur<getop(mem,p),s}
        (cpart2(getop(mem,p),ac) *
          dpart(p,ac,arg,getop(mem,p)))
    + if eqs(cu,getop(mem,p)) then
      {cntl<opfetch,s}{s}(cpart1(cu,ac) *
          dpart(p,ac,arg,getop(mem,p)))
```

Abstraction on *cur* and *cntl* leads to:

```
cpart(cu,ac) * dpart(p,ac,arg,cu)
      - cntl - cur =
    if noteq(cu,getop(mem,p)) then
      {s}{s}(cpart2(getop(mem,p),ac) *
              dpart(p,ac,arg,getop(mem,p)))
    + if eqs(cu,getop(mem,p)) then
      {s}{s}(cpart1(cu,ac) *
              dpart(p,ac,arg,getop(mem,p)))
```

which simplifies to

```
{s}{s}(cpart2(getop(mem,p),ac) *
        dpart(p,ac,arg,getop(mem,p)))
```

If $\text{IMP}k(\text{cu},\text{p},\text{ac},\text{arg},\text{n})$ is defined as

```
(cpartk(cu,ac) *
dpart(p,ac,arg,cu) - cntl - cur) *
DEV(n) * newcomputer(p,ac)
```

then further expansion of the composition operator gives

```
IMP(cu,p,ac,arg,n) <=
    if and(noteq(n,1),noteq(decr(n),1)) then
    {s}{s}
    IMP2(getop(mem,p),p,ac,arg,decr(decr(n)))
```

The behaviour of `SPEC` is given by

```
SPEC(p,ac,n) <= if eqs(n,1) then
      (if eqs(getop(mem,p),ADD) then
          {t,s,acc<add16(ac,getarg(mem,p)),
                            pc<incr(p)}
          SPEC(incr(p),
                 add16(ac,getarg(mem,p)),M)
     + if eqs(getop(mem,p),SUB) then
          {t,s,acc<sub16(ac,getarg(mem,p)),
                            pc<incr(p)}
          SPEC(incr(p),
                 sub16(ac,getarg(mem,p)),M)
    + if eqs(getop(mem,p),LD) then
          {t,s,acc<getarg(mem,p),
                            pc<incr(p)}
          SPEC(incr(p),getarg(mem,p),M)
   + if and(eqs(getop(mem,p),JZRO),eqs(ac,0))
          then {t,s,pc<getadd(mem,p)}
          SPEC(getadd(mem,p),0,M)
    + if and(eqs(getop(mem,p),JZRO),
                    noteq(ac,0)) then
      {t,s,pc<incr(p)}SPEC(incr(p),ac,M))
   + if noteq(n,1) then {s} SPEC(p,ac,decr(n))
```

Now it can be seen that the two behaviours are equivalent at the first level (capable of performing the same initial actions) if `noteq(n,1)` is true, since in this case both behaviours can simply accept an s event.

Expansion of the composition and abstraction operators for `IMP2` gives

```
IMP2(cu,p,ac,arg,n) <=
   if and(or(eqs(cu,ADD),or(eqs(cu,SUB),
      eqs(cu,LD))),noteq(n,1)) then
     {s}IMP3(cu,p,ac,getarg(mem,p),decr(n))
  + if and(and(and(eqs(cu,JZRO),noteq(n,1)),
      eqs(decr(n),1)),eqs(getop(mem,p),JZRO))
      then
    (if and(noteq(getadd(mem,p),p),eqs(ac,0))
        then {s}{pc<getadd(mem,p),s,t}
           IMP(JZRO,getadd(mem,p),ac,arg,M)
   + if noteq(ac,0) then {s}{pc<incr(p),s,t}
           IMP(JZRO,incr(p),ac,arg,M))
```

In the last two branches of this behaviour, events involving external ports other than *s* can at last be observed. The process of showing that the actions of IMP and SPEC are the same now becomes a little more involved. It was shown above that the behaviour of IMP was given by

```
IMP(cu,p,ac,arg,n) <=
   if and(noteq(n,1),noteq(decr(n),1)) then
   {s}{s}
   IMP2(getop(mem,p),p,ac,arg,decr(decr(n)))
```

The behaviour of `IMP2` can be inserted in this expression, with `getop
(mem,p)` replacing `cu` and `decr(decr(n))` replacing n. This gives:

```
IMP(cu,p,ac,arg,n) <=
   if and(noteq(n,1),noteq(decr(n),1)) then
    (if and(or(eqs(getop(mem,p),ADD),
         or(eqs(getop(mem,p),SUB),
         eqs(getop(mem,p),LD))),
         noteq(decr(decr(n)),1)) then
      {s}{s}{s}IMP3(getop(mem,p),p,ac,
          getarg(mem,p),decr(decr(decr(n))))
   + if and(and(and(eqs(getop(mem,p),JZRO),
         noteq(decr(decr(n)),1)),
         eqs(decr(decr(decr(n))),1)),
         eqs(getop(mem,p),JZRO)) then
    (if and(noteq(getadd(mem,p),p),
         eqs(ac,0)) then
      {s}{s}{s}{pc<getadd(mem,p),s,t}
          IMP(JZRO,getadd(mem,p),ac,arg,M)
   + if noteq(ac,0) then
      {s}{s}{s}{pc<incr(p),s,t}
          IMP(JZRO,incr(p),ac,arg,M)))
```

In every branch, the first 3 guards are just s ticks. It is easy to see that
this is the same behaviour that would be exhibited by SPEC if the initial value of n were greater than 3. The next action of SPEC would be
`{pc<getadd(mem,p),s,t}` if the conditional `and(eqs(getop
(mem,p),JZRO),eqs(ac,0))` were true and n (which has by this
point been decremented 3 times) equal to 1, i.e. the initial value of n must
be 4. If this is the case, then all the predicates in IMP's behaviour that
involve n evaluate to true. Finally, if `and(eqs(getop(mem,p),
JZRO),eqs(ac,0))` is true and `noteq(getadd(mem,p),p)` is
also true then the fourth action of IMP will also be `{pc<getadd(mem,
p),s,t}`. This extra condition arises because the descriptions of com-
ponents of the implementation were written to preclude fictitious events,
using the techniques of Section 3.2, while the specification of newcomputer was not. The significance of this difference is discussed below.
Aside from that, however, it has been successfully shown that the implementation behaves correctly with respect to the specification for the

performance of the JZRO instruction when the accumulator contains 0. A similar approach to that just described could be adopted to establish that the implementation executes JZRO correctly when the accumulator does not contain 0, and also that it performs the other three instructions as required by the specification. It should be noted that the initial value of n may vary from one instruction to another, but that this does not invalidate the result of the verification.

6.5 *Discussion*

In this Chapter, the tasks of specification, design and validation by formal means have been illustrated with an example of realistic size. This final Section discusses some of the issues that have been raised by this exercise.

First of all, it has been demonstrated that the descriptive power of enhanced CIRCAL is sufficient to allow quite complex devices to be specified. By allowing variables of any type to be used as state parameters and passed on ports, and by giving access to the function and type definition capabilities of a language such as ML, it is possible to produce quite succinct and intelligible descriptions of devices at high levels of abstraction.

In the design task, the benefits of using small design steps (i.e. adding a small amount of structural information at each step) were demonstrated. Hierarchical design is the application of the well-established technique of problem reduction to design, and it follows that the benefits of this technique are lost if the distance between levels is too large. Furthermore, the ability of a language to support closely-spaced levels was illustrated by the use of the datatype `cntltype`, which enabled the descriptions of `cpart` and `dpart` to be more abstract.

The validation task is also simplified by the use of fine design steps, because the differences between the behaviour of the specification and of the implementation are reduced. In this example, the main difference in behaviour that had to be reconciled was the different granularities of time used at the two levels of abstraction. This was shown to be a fairly straightforward task, demonstrating both the suitability of CIRCAL for dealing with the analysis of timing characteristics and the feasibility of comparing behavioural descriptions with different amounts of timing information.

One unsatisfactory aspect of the verification was the fact that an additional predicate was present in the behaviour of the implementation. This arose from the different modelling philosophy adopted at the implementation level, which ensured that fictitious events could not occur. Since such events had not been precluded in the specification, it came as no surprise to discover this discrepancy when the verification was attempted. In this example, it was easy to spot the cause of the discrepancy and dismiss it; in many situations, the reason for such a difference might be more obscure, and a designer may only be able to conclude that the implementation is incorrect. This illustrates the point, made in Chapter 3, that a consistent modelling philosophy should be adopted throughout a design.

Working through this example has shown that the tasks of specification, design and validation are not independent: the way in which each is approached affects the ease with which the others can be tackled, and all three tasks are influenced by the characteristics of the language that is used. This illustrates the importance of developing a language in the context of an integrated approach to design and validation.

SEVEN

____

CONSTRAINTS

____

In the foregoing discussion, there have been several occasions on which
the word 'constraints' has been mentioned. This word is frequently used
both by designers and researchers in the field of design and validation, but
it is rarely defined. The investigation of the use of constraints in design
and validation has shown that the word can be used to embody several
different concepts but that there is a quite specific definition which is particularly useful. The reasoning that underlies the choice of this definition
is explained below.

Constraints are generally quite extensively used by designers, as some
of the following examples will illustrate. They are normally used, however, only in an informal way. From the large body of work that has recently been undertaken on this topic [Davie 88, Herbert 88, Langevin 88,
Milne 88a, Subramanyam 88], it has become apparent that a formal treatment of constraints may make a significant contribution in the tasks of
specification, design and validation. In this Chapter, the ways in which
formally specified constraints may provide assistance in each of these
tasks will be examined in detail. In addition, the effect of using constraints in one task on the execution of other tasks will be presented. The
fact that constraints can only be formally used in a formal, language-
based methodology strengthens the argument for the approach proposed
in the preceding chapters. The first step towards making use of constraints is to formulate a precise definition; to this end, some common
examples are considered.

7.1 Introductory Examples and Definitions

In the most general sense, a constraint is a collection of statements that
impose restrictions on something, i.e. place limits on what that 'something' can do. Some familiar concepts in circuit design and validation
that fall within the scope of this definition are:

1. setup, hold and rise times for a D flip flop;

2. temporal relationships between clock phases and data input ports in a PLA controlled by a two phase clock;
3. the restriction on an RS flip flop that its two inputs may not be simultaneously high;
4. geometric design rules;
5. the choice of a certain design style (e.g. microprocessor / microcode engine / PLA), thereby restricting future design decisions;
6. the use of a certain clocking scheme.

It is worth considering what is actually constrained, and in what way, in each of the above examples. In the first, the devices that provide the input signals to the flip flop are constrained; for example, the value on the clock port must change from false to true in a period that is less than the specified rise time. The setup and hold times restrict the times at which the values on the data port may change relative to changes on the clock port. Similarly, in the second example it is the devices that are to be connected to the PLA which are constrained. The two clock phases must be generated without overlap and with suitably long pulse widths and separations, and the changes on the input ports must occur only at certain times in the clock cycle. The third example differs slightly, in that it is the *values* on ports which are restricted rather than the times of changes, but it is still the devices connected to the flip flop which are being constrained.

The common thread in these first three examples is that in each case the constraint associated with a device restricts the behaviour of other devices connected to it. These devices may be called the *environment* or *context* of the first device, and this type of constraint is therefore referred to as a contextual constraint. The two words 'environment' and 'context' will be used interchangeably to refer to the set of devices to which a device is connected. (The concept of a device's environment was first introduced in Section 2.3.1; a typical environment is pictured in Figure 2.2.) A contextual constraint, which is associated with a single component, is therefore some set of restrictions on the behaviour of the devices that are connected to that component.

Usually these restrictions will relate to the *times* at which new values may be placed on the device's inputs by the environment, or the actual *values* that may be supplied. Thus it will often be the case that the only part of the environment that is of interest is the part that is connected to the inputs of a device. There are exceptions to this, however; for example, the acceptable time for providing a new input value may be dependent on the time at which the last output change occurred.

The remaining examples of constraints involve a different type of restriction. The geometric design rules constrain a designer; they constitute a set of restrictions on the ways in which he may arrange polygons in the various layers. Similarly, in selecting a particular design style at some point in the design process, the designer places restrictions on himself which reduce the range of design options open to him at subsequent stages of the design. The use of a particular clocking scheme also restricts the designer, who is then required to design all components such that they will function and interact correctly under that scheme.

These three constraints are thus seen to be in a different category from the first three, in that, rather than restricting the behaviour of a component or set of components, they all restrict the choices of the designer. These constraints are very difficult to treat in a formal way; of the examples above, only the geometric design rules can be easily expressed as an unambiguous set of rules. (A somewhat formal approach to restricting the designer, however, is described in Section 4.1.2.) By contrast, the former type of constraint can readily be described in a formal way, as illustrated below. Thus, although both types of constraint may have a role to play in a design methodology, the remainder of this Chapter will concentrate on contextual constraints and the ways in which their formal description and use may serve the design and validation process.

A formal treatment of contextual constraints can be of assistance in each of the main tasks of the methodology proposed in Chapter 1; later sections will discuss the roles of constraints in the tasks of specification, design and verification and examples of constraints that can be used in each task will be provided. Before making use of constraints, however, it is necessary to have a formal way to describe them, and this is presented in the next Section.

7.2   *Specification of Constraints*

A contextual constraint was defined above as a restriction on the behaviour, in terms of the events that may take place and the times when they may occur, of a device's environment. The environment consists of a collection of components connected to the device with which the constraint is associated, so there is no reason not to consider it as just another device.

In previous discussion of CIRCAL it was noted that it is all too easy to write descriptions in which input events were not allowed at certain times. In the description of real hardware this was considered to be a disadvantage, but in the specification of constraints the ability to prohibit certain events from occurring is just what is required. It therefore seems reasonable to suppose that a formal specification of a constraint for a device could be achieved by writing a CIRCAL description of some abstract component representing the environment of the device. It was seen in Chapter 3 that CIRCAL specifications could be constructed from a number of parts, so that this description might be considered a partial specification of the environment. A few examples will serve to illustrate how constraints may be specified in this way.

Example

To specify the constraint on an RS flip flop that its two inputs may never simultaneously be set to the value true, an abstract box called CON, with ports *r* and *s* could be specified as follows:

```
CON <= {r<true}{r<false}CON
     + {s<true}{s<false}CON
```

Explanation

A `true` event on either port can only be followed by a `false` event on the same port; this ensures that the two ports are never both set to the value `true` simultaneously. It should also be noted that this constraint specification places no other restrictions on the behaviour at these ports.

Constraints often occur in the form of acceptable limits within which some behavioural parameter must lie. One of the initial examples, a latch's setup time, falls into this class.

Example

For a setup time of *n* time units, the constraint is that a waveform must assume a steady value *at least n* units before the latching edge of the clock. This type of non-rigid description can be achieved in CIRCAL using a technique similar to that which was described in Section 3.2.4.

```
TGEN <=   {t}TGEN
   + {t1,t}{t}{t}{t}{t2,t}TGEN
CONA <= {data>x}{t1}{clk,t2}CONA
CON <= TGEN * CONA - t1 - t2
```

Explanation

This constraint uses a constructive specification technique, the constraint box CON being constructed from two simpler boxes. The box TGEN may either passively accept ticks, or on the occurrence of a pulse on *t1* will wait for 4 more ticks before generating a pulse on *t2*. The box CONA, after accepting an input event on *data*, generates the pulse on *t1*. No event may then happen on *clk* until the pulse on *t2* occurs. This ensures that a `clk` event cannot occur less than 4 ticks after an event on the *data* port. The relationship between the various events is illustrated in Figure 7.1.

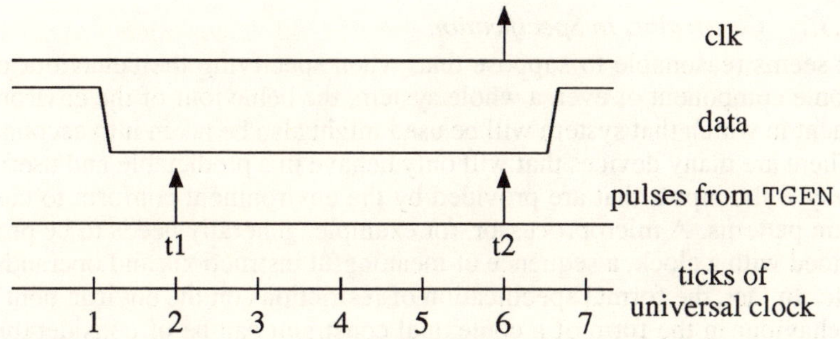

Figure 7.1: Timing diagram for setup time constraint

Example
The third of the examples of contextual constraints mentioned above
could be described in the following way:

```
CON <= {phione<true}{phione<false}CON
       + {data>x}CON
```

Explanation
The sort of this abstract box is *{phione,data}*. The second branch of
its description allows any events on the *data* port. However, if the value
on the input-latching clock *phione* becomes true, the only event that can
follow is {phione<false}, i.e. there may be no change on the *data*
port until the clock is false again.

From these examples it is apparent that CIRCAL can be applied to
the task of constraint description by writing behavioural descriptions in
which certain events are disallowed. In Section 7.4 the problem of describing constraints in the high level language SuperC will be discussed.
The following Section shows how formally specified constraints may be
put to use in all phases of the design and verification process.

7.3   The Uses of Constraints

Having seen a few examples of constraints that are commonly used —
although only in an informal way by most designers — and having established a mechanism by which such contextual information can be formally represented, it remains to be shown:

(a) that there are more than a few isolated situations in which contextual
constraints can be used; and
(b) that a formal treatment of such constraints will provide assistance
in design and validation.

The following Sections will serve to demonstrate both these points.

7.3.1   *Constraints in Specification*

It seems reasonable to suppose that, when specifying the behaviour of some component or even a whole system, the behaviour of the environment in which that system will be used might also be taken into account. There are many devices that will only behave in a predictable and useful way if the inputs that are provided by the environment conform to certain patterns. A microprocessor, for example, generally needs to be provided with a clock, a sequence of meaningful instructions and operands, etc. In fact, the formal specification of restrictions on the environment's behaviour in the form of a contextual constraint can be of considerable assistance in the specification task, as will become apparent in the following discussion.

In previous chapters the problems of writing specifications that are suitable for subsequent use in the validation task have been discussed. It has been shown that some approaches to specification which appear to simplify the task can actually prevent the successful validation of a implementation which, had it been correctly specified, would have been satisfactory. This sort of problem arose because the specification was written under some assumptions about the behaviour of the environment of the device being specified; when these assumptions turned out to be invalid, the validation of the design could not be carried out successfully. This is an unsatisfactory situation, as it leads to the false conclusion that an implementation is incorrect, when in fact it is the specification technique that is to blame. Contextual constraints can resolve this problem, as the following example, an inverter with delay, illustrates.

Example

In Section 5.2.2 a specification technique was used in which output events lagged input changes by some time interval and no input changes were allowed during this time. In writing a specification in this way, the implicit assumption is that the environment of the specified device will not fire input changes at it 'too rapidly', i.e. an input event will be followed by a sufficiently long delay to allow an output event to occur. However, this assumption was never made explicit, and the result was that the verification failed.

The solution to this problem is to associate a constraint with the specification, as follows:

```
INV(x) <=
     {in>y:noteq(y,x),t}{out<not(y),t}INV(y)
   + {t}INV(x)
CON <= {in>x,t}{t}CON + {t}CON
```

Explanation

The first line states that a new value y can be input on one tick and that the complement of that value will be output on the next tick of the universal clock. The abstract device CON represents the contextual constraint. The sort of this device is {*in*, *t*}. If an event occurs on the port *in* on one tick, then the only possible guard to follow this is {t}. Thus, no input event can follow on the next tick.

Suppose now that the inverter is to be connected to a component which will generate events on the port *in* on successive ticks of the universal clock, such as the following:

```
DEV <= {in<true,t}{in<false,t}DEV
```

In order to check whether the assumptions about the environment's behaviour were correct, it is necessary to find whether DEV *satisfies* the constraint CON in some sense. What is required is that DEV does not try to perform any actions that are not permitted by CON. Now if DEV did contain some actions that were prohibited by CON, then DEV * CON would not contain these actions. Therefore, DEV * CON would not be equal to DEV. Provided that the sorts of DEV and CON are the same, DEV * CON will certainly not have any more actions than DEV. So, the constraint is satisfied if

```
DEV * CON  = DEV
```

where = means 'can perform exactly the same actions as'. This may be contrasted with the satisfaction requirement for implementations described in Section 5.2.1 [Milne 85b]. An implementation satisfies its specification if it can perform *at least* all the actions of the specification. An environment satisfies its constraint if it can perform *no more* actions than the constraint allows. This can readily be tested:

```
DEV * CON  =
    ({in<true,t}{in<false,t}DEV) *
    ({in>x,t}{t}CON)
 = {in<true,t}(({in<false,t}DEV) * ({t}CON))
 = {in<true,t} /\
```

which is clearly not equal to DEV, which is a non-terminating behaviour. However, if the inverter is connected to a different device, say DEV1, which waits one tick between events on *in*, specified as

```
DEV1 = {in<true,t}{t}{in<false,t}{t}DEV1
```

then a check on the satisfaction of the constraint proceeds as follows:

```
DEV1 * CON  =
    ({in<true,t}{t}{in<false,t}{t}DEV1) *
    ({in>x,t}{t}CON)
  = {in<true,t}(({t}{in<false,t}{t}DEV1) *
    ({t}CON))
  = {in<true,t}{t}(({in<false,t}{t}DEV1) * CON)
  = {in<true,t}{t}{in<false,t}{t}(DEV1 * CON)
```

It can be seen that this behaviour is equivalent to DEV1, so DEV1 satisfies the constraint CON.

Full and Partial Specifications
One question to be addressed here is this: when writing a specification, how does the designer know whether it is necessary to write a contextual constraint to ensure that the environment in which the part is to be used will not violate the assumptions under which it was specified? The key to the answer to this question lies in the concepts of full and partial specifications.

A specification may be classed as 'full' if, in any state, any combination of valid events on the input ports is possible. 'Valid' events are those which represent a genuine change in value, as discussed in Section 5.2.2. This means that for a device with n ports on which inputs can occur, there will be at least $2^n - 1$ guards, to allow for any possible combination of input events. If an input event has a qualifying predicate that is more restrictive than one to ensure the validity of the event, then there must be another guard to take account of the cases in which that predicate is not satisfied. Furthermore, if a state is parameterised, there must be a full complement of input events for any possible set of state parameter values. It should also be noted that a state need not be named explicitly; any part of a behavioural description that follows a guard can be considered a state. A full specification can accept any input event at any time. It therefore does not require a contextual constraint, since its environment cannot possibly attempt to perform an input event that it would not accept.

Anything that does not conform to the definition of a full specification is a partial specification. The following examples illustrate both classes.

Example
A full specification of an and gate with no delay could be written as:

```
AND(p,q) <= if p then
      {b>x:noteq(x,q),out<x}AND(p,x)
  + if q then {a>x:noteq(x,p),out<x}AND(x,q)
```

```
 + if eqs(p,q) then
      {a>x:noteq(x,p),b<x,out<x}AND(x,x)
 + if not(p) then {b>x:noteq(x,q)}AND(p,x)
 + if not(q) then {a>x:noteq(x,p)}AND(x,q)
 + if noteq(p,q) then
      {a>x:noteq(x,p),b>y:noteq(y,q)}AND(x,y)
```

Explanation

The first three lines of the description take account of input events that
produce an output change, the leading conditional ensuring that a valid
output event will result. Conversely, the last three lines take account of
input events that will produce no output change. The important point
here is that for any values of p and q, input events may take place on *a*,
b or both. Thus, all possibilities are taken into account, and the specification is a full one.

Example

The following is a full specification of an RS flip flop:

```
RSFF(x,y,q) <= if and(not(x),q) then
     {r<true,t}RSFF1(true,y,q)
 + if and(not(x),not(q)) then
      {r<true,t}RSFF(true,y,q)
 + if and(not(y),not(q)) then
      {s<true,t}RSFF1(x,true,q)
 + if and(not(y),q) then
      {s<true,t}RSFF(x,true,q)
 + if x then {r<false,t}RSFF(false,y,q)
 + if y then {s<false,t}RSFF(x,false,q)
 + if eqs(x,y) then
     {r>p:noteq(x,p),s<p,t}RSFF(p,p,q)
 + if noteq(x,y) then
     {r>p:noteq(x,p),s<not(p),t}
         RSFF1(p,not(p),q)
 + {t}RSFF(x,y,q)
RSFF1(x,y,q) <=
   {t,out<not(q)}RSFF(x,y,not(q))
 + if and(x,not(y)) then
      ({r<false,t,out<not(q)}
         RSFF(false,y,not(q))
    + {s<true,t,out<not(q)}
         RSFF(x,true,not(q)))
 + if and(not(x),y) then
```

```
       ({s<false,t,out<not(q)}
           RSFF(x,false,not(q))
        + {r<true,t,out<not(q)}
           RSFF(true,y,not(q)))
   + if noteq(x,y) then
       {r>p:noteq(x,p),s<not(p),t}
           RSFF(p,not(p),q)
   + if eqs(x,y) then
       ({r>p:noteq(x,p),s<p,t}RSFF(p,p,q)
        + {r>p:noteq(x,p),t}RSFF(p,y,q)
        + {s>p:noteq(y,p),t}RSFF(x,p,q))
```

Explanation

In both state definitions, the parameters x, y and q represent the values on the *r* (reset), *s* (set) and *out* (output) ports respectively. The state RSFF(x,y,q) represents the stable states, in which no output change is due to occur. This is evident from the last branch in the description. In certain circumstances, an input event leads to the unstable states denoted by RSFF1(x,y,q). In these states, if there is no new input on the next tick of the universal clock, then the value on the output is complemented, as indicated by the first branch in the description.

In the unstable states, if the next tick is accompanied by another input change, then this may cancel the pending output event (by leading back to a stable state) or it may have no effect.

The assertion that this is a full specification, according to the above definition, can be tested. The specification is full if, in all possible states of the device, it is possible for any combination of valid input events to occur. In any state, there are three possible combinations of valid input events: a change on the reset port, a change on the set port, and a simultaneous change on both ports.

A change on the reset port can take place from a state RSFF(x,y,q) under the following circumstances:

- when and(not(x),q) is true;
- when and(not(x),not(q)) is true;
- when x is true.

Given that $(\text{and(not(x),q)} \wedge \text{and(not(x),not(q))} \wedge x) = \text{true}$, it can be concluded that a change on the reset port is always possible. Similar checks for changes on the set port only and on both ports at once show that, for any values of x, y and q, all three combinations of input events are possible.

To ensure that the specification is full for the states denoted by RSFF1 (x,y,q) it is necessary to carry out the same checks for each of the possible combinations of valid input events. Thus the same conclusion, that each of the three combinations can occur for any values of x, y and q, is reached.

This process of checking, as presented above, seems fairly laborious. However, it would be reasonably straightforward to automate it, since it is a purely mechanical process.

Before discussing the problems encountered in formulating a full specification for this device, it is helpful to consider how a partial specification of the same device might be written. If it is assumed that *s* and *r* will never be simultaneously set to true, then the following partial specification can be used:

```
RSFF(x,y,q) <= if and(not(x),q) then
      {r<true,t}RSFF1(true,y,q)
 + if and(not(x),not(q)) then
      {r<true,t}RSFF(true,y,q)
 + if and(not(y),not(q)) then
      {s<true,t}RSFF1(x,true,q)
 + if and(not(y),q) then
      {s<true,t}RSFF(x,true,q)
 + if x then {r<false,t}RSFF(false,y,q)
 + if y then {s<false,t}RSFF(x,false,q)
 + if noteq(x,y) then
     {r>p:noteq(x,p),s<not(p),t}
         RSFF1(p,not(p),q)
 + {t}RSFF(x,y,q)
RSFF1(x,y,q) <= {t,out<not(q)}RSFF(x,y,not(q))
 + if x then
     {r<false,t,out<false}RSFF(false,y,false)
 + if y then
     {s<false,t,out<true}RSFF(x,false,false)
 + {r>p:noteq(x,p),s<not(p),t}RSFF(p,not(p),q)
```

Explanation

As before, RSFF(x,y,q) represents the stable states and RSFF1(x, y,q) the states in which an output is pending. In this description, however, the possibility of ports *r* and *s* being set to true simultaneously is not taken into account. The specification is therefore a partial one. For example, in a stable state, simultaneous events cannot take place on the two inputs unless noteq(x,y) is true (from the penultimate branch of the definition of RSFF(x,y,q)).

By assuming that the two inputs will never both be set to true at the same time, the process of writing the specification becomes much simpler. In particular, the possible scenarios in the unstable states denoted by RSFF1(x,y, q) are much reduced in both number and complexity. For example, it can now be assumed that RSFF1(x,y,q) will only be reached if exactly one of x and y is true, so there is no need to write a specification that takes account of any other possibilities. The specification that results is not just shorter, it is much easier to formulate.

In summary, the disadvantages of full specifications are:

- The number of possible events that must be taken account of in each state is larger than it would be if a partial specification were written, the difference being quite significant in some situations;
- Making sure that all possible event combinations have been covered is essential and may be quite a laborious task, especially when conditionals and input parameter predicates are used;
- Deciding how the device should behave under certain circumstances can be difficult and needlessly time-consuming. For example, what should the behaviour of the RS flip flop be when a reset event is immediately followed by a set event?

The advantages of writing partial specifications which only describe that part of the device's behaviour that is of interest can thus be clearly seen. In writing a partial specification, however, one is assuming that certain patterns of inputs will not occur. In order to ensure that this assumption is justified, there must be a constraint on the environment of the partially specified device which prohibits those patterns of inputs. It may then be necessary to check that the environment meets the constraint, or to design it with the constraint information taken into account. The consequences of designing a device's environment without regard for the assumptions under which that device was specified have already been demonstrated in Section 5.2.2. Thus the use of constraints in one task starts to influence other tasks. The spread of constraints in this way is discussed in the following Section.

One issue that has been ignored so far is the question of whether a constraint is adequate to prevent events that a partially specified device cannot accept. Just as a full specification must accept all possible valid inputs, a partial specification must be able to accept all inputs that its associated constraint does not preclude. A way in which this can be guaranteed is presented in Section 7.4.

At the start of this Section the question was posed whether contextual constraints could only be applied in a few isolated situations. This question can now be partially answered. The ability to write a partial specification is something that may be required in the description of almost any device. Contextual constraints must be written if partial specifications are to be safely used. The role of constraints in the specification task may therefore be quite significant and extensive.

7.3.2   Constraints in Design

It has been shown that specifications may be more easily written if a contextual constraint is attached to that specification. The design task consists of partitioning a box, which has some specification, into smaller cells, and assigning specifications to those cells. This suggests two ways in which constraints may play a part in the design task:

1. in the designing of a box whose specification, because it is partial, has an associated constraint;
2. in the specification of cells that constitute the environment of a device that has an associated constraint.

Designing a Partially Specified Device

To examine how the association of a constraint with a box's specification might affect its design, it is worth returning to the example of the RS flip flop described above. When this device was specified fully and without a constraint, it was necessary to decide how it would behave under various unusual conditions such as the arrival of a set pulse when the value on the reset port was already true. Because some response had to be specified under all of these conditions, even though the designer may not actually care what happens, the problem of producing a suitable implementation becomes significantly harder. For the implementation must perform at least all the actions that the specification can perform; if some of the actions of the specification were specified needlessly and arbitrarily, a considerable amount of effort may be expended in constructing an implementation that performs the same arbitrary actions.

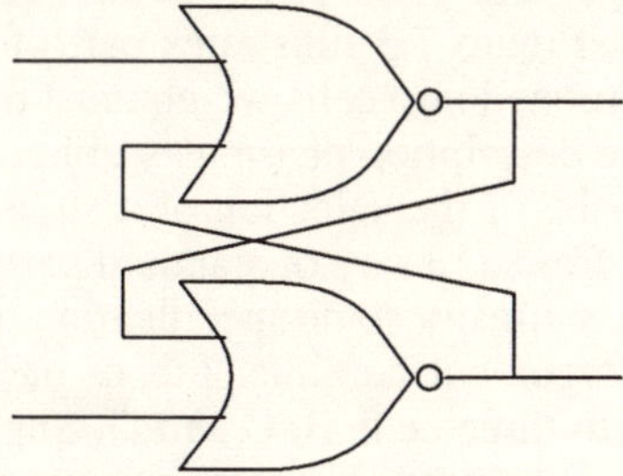

Figure 7.2: NOR gate implementation of RSFF

In the flip flop example, the specification is written in such a way that if the value on r is true, and the event s<true occurs, then the value on

the output will be complemented on the next tick. Now, if the common cross-coupled nor gate implementation of the RS flip flop, pictured in Figure 7.2, is used, it will actually fail to meet the specification because it does not behave that way. The same implementation could, however, be used to implement the partial specification of the flip flop, because it behaves correctly under all the input stimuli which that specification allows. It will also accept input stimuli other than those covered by the partial specification, but this does not make it an unsatisfactory implementation; it simply has a richer behaviour than the specification. Some further consequences of this difference in behaviour between implementation and specification are discussed in Section 7.3.3. The important point to note here is that a simple and quite acceptable implementation could actually be rejected because the specification was unnecessarily stringent.

Thus there is another argument for the use of partial specifications with associated constraints. Such specifications are not just easier to write. By specifying only those aspects of a behaviour that actually matter, the construction of a satisfactory implementation is also made easier. Therefore, it may be desirable to specify a device partially whenever it is known that its environment will only provide a limited set of stimuli, thus providing the designer with more flexibility in choosing an implementation. This particular use of contextual information is the subject of some current research [Subramanyam 88].

Designing Environments Under Constraints

The second role of constraints in the design task is in the designing of environments for devices whose specifications have associated constraints. This use of constraints may be most easily explained by way of an example.

In Chapter 4, design was seen to consist of two tasks, partitioning and description. Here again, Figure 7.3 illustrates part of a typical design step. Box A is to be partitioned into cells which must then be described. It will be recalled that the description phase may either be performed by the designer or may consist of the selection of a suitable part, with an associated specification, from a library of standard parts.

In this example, there is already some specification for A, which may for the purposes of the argument be either full or partial. It has been decided to partition A into three cells B, C and D. Suppose that the description phase begins with the writing of a specification of the cell B and that for some reason this specification has an associated contextual constraint. This may be because B has been partially specified for one of the reasons discussed above, or it may be that B is a part taken from a library of standard parts, which has an attached constraint such as setup and hold

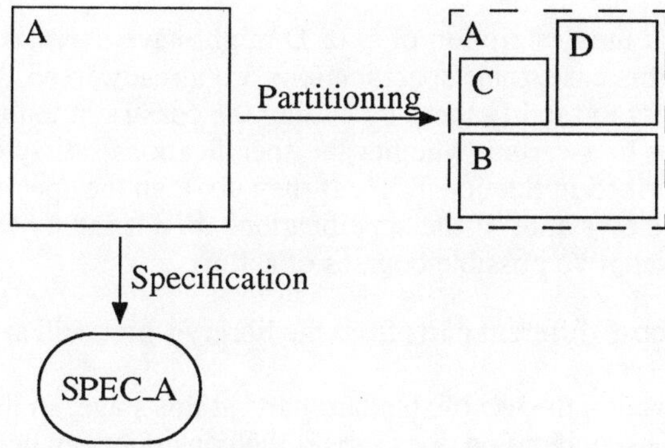

Figure 7.3: A box to be designed using constraints

time requirements. This constraint constitutes a statement about the allowed behaviour of the environment of B. The cells C and D are a part of the immediate environment of B and must therefore be behaviourally described in such a way that they conform to the environmental constraint.

It was seen in Section 3.2.2 that behavioural descriptions in CIRCAL may be built up from parts by the technique of constructive specification. This technique may be effectively used in this situation; the first part of the specification for each of the boxes C and D is simply the contextual constraint associated with B. If this constraint is denoted by CON_B, then a specification for C may be written as

```
C <= CON_B * SPEC_C
```

and SPEC_C may be written without any concern for the constraint; constructed in this way, C will be guaranteed to satisfy the constraint CON_B. If C also has an associated contextual constraint, denoted by CON_C, then D would be specified as

```
D <= CON_B * CON_C * SPEC_D
```

In the situation just described, constraints from neighbouring parts were used to construct the specifications of devices which were unspecified at the time the constraint was written. That is, when CON_B was written, there was as yet no specification for the parts C and D. If such specifications had already existed, it would have been necessary to modify them to ensure that they would satisfy CON_B. If the specifications had been written by the designer, this would present no problem; it would be a simple matter of composing CON_B with SPEC_C and SPEC_D respectively as before. The one slightly more awkward situation that could arise is that the description of C or D might have been taken from a library. In this case, their specifications are already fixed, so that one is not at liberty to modify them by adding the constraint to them. The first step then is to ascertain whether the specifications satisfy CON_B in the sense described in Section 7.3.1. If they do, then the specifications may be left as they stand. If the specifications do not satisfy the constraint, then there are two possible courses of action:

- to choose different parts from the library which will satisfy the constraint; or
- to abandon the use of standard parts at this stage, so that a new part whose specification does satisfy the constraint can be designed.

If the second option is chosen, then the components C and D will be treated as the specifications that are to be implemented at the next level down in the design hierarchy.

In the preceding discussion, the environment of B was assumed to involve only the parts C and D. This is actually something of an over-simplification, as the environment of a part is actually the sum total of all parts that are immediately connected to it. If the 'parent' part A, which is to be implemented by parts B, C and D, is actually part of some larger system, then it is entirely possible that part of the environment of B will lie outside of A. This situation is illustrated in Figure 7.4, in which A is connected to another box, called X.

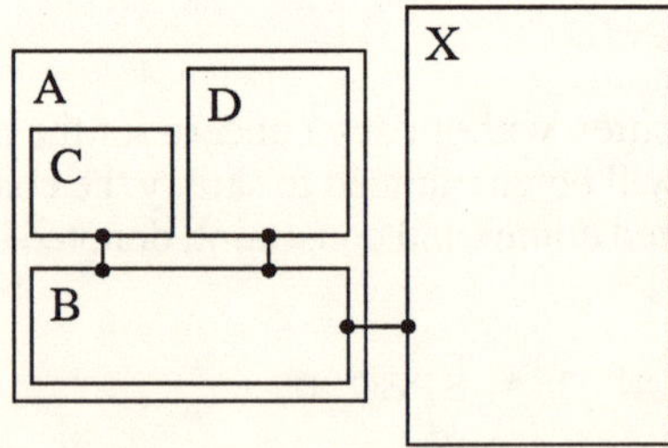

Figure 7.4: An environment extending beyond the parent box

This situation is rather more difficult to deal with. In a top-down approach to design, all the parts to which A is connected would have been at least specified, and possibly implemented as well, before A is designed. Therefore, it is not generally possible to impose the constraints associated with B on those parts of its environment that lie outside A. The first step is to test whether the constraints associated with B are satisfied by those parts of its environment. If they are, there is no problem. If not, there are two options:

- to take steps to modify the constraint on B so that it will be satisfied by the environment as it stands. This may involve modification of the specification of B, or the use of a different part from the library;
- to modify the environment so that it satisfies the constraints.

Modification of the environment means that the design is no longer rigorously top-down, but iterative. In real design situations, iterative design is actually the more common approach.

It is noteworthy that the uses of constraints just described have not been motivated by a desire to ease the task at hand. Instead, constraints are used here because they have been generated at some other stage and it is now necessary to ensure that they are satisfied by the environments to which they apply. Thus, it is at this stage that constraints may actually make the design task more difficult rather than less. However, since the effect of constraints in this situation may be to limit the choices available to the designer, they may actually be considered a design aid. At any rate, whether constraints hinder or aid the designer in this task, their use has already been seen to be important in other tasks. One task, validation, remains to be discussed; this is done below.

7.3.3 Constraints in Verification

The formal treatment of constraints is of greater use and importance when the method of validation is also formal. This Section therefore concentrates on formal verification rather than simulation. There are two ways in which constraints are involved in the verification task:

- the verification of a device may be simplified if it is only partially specified, requiring a constraint to be attached;
- when a partially specified device is verified, it may be necessary to apply an additional constraint to its environment to ensure that an implementation that satisfies the partial specification will function correctly.

These two roles of constraints in verification are discussed in detail below.

Constraints to Ease the Verification Task

It has already been seen that it may be more difficult to implement a system if its specification contains more information than is absolutely necessary. The use of partial specifications is therefore an aid to the design task. It is reasonable to suppose that the same may be true of the verification task, for the following reason: if a system is only partially specified, then the range of input stimuli that it can accept is reduced. Therefore, the amount of work that has to be done to show that the implementation responds to all possible input stimuli in the same way as the specification will hopefully also be reduced. The following example will show how this reduction in verification effort may be achieved.

Example

A piece of combinational logic is 'sandwiched' between a pair of latches as illustrated in Figure 7.5.   A partial specification for such a device

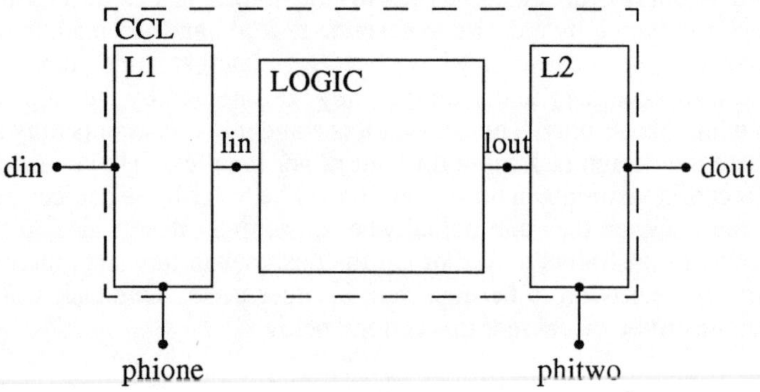

Figure 7.5: Clocked combinational logic

could be written under the assumptions that the two clock phases are non-overlapping and separated by a certain interval, and that the input lines will not be subject to changes during the time that *phione* is true. In CIRCAL these constraints could be described as follows:

```
CON  <= {phione<true}{phione<false}CONA
  + {din>x}CON
CONA <= {phitwo<true}CONB + {din>x}CONA
CONB <= {phitwo<false}CON + {din>x}CONB
```

Explanation

In the initial state CON, a new data value may be input, or the current value may be latched by *phione*. Once *phione* becomes true, the only thing that may happen next is that it may become false again; no changes on the input port *din* or the other clock phase are possible. After *phione* becomes false, *phitwo* is re-enabled, as is *din*. A rising edge on *phitwo* leads to state CONB, in which changes on the input are still possible, and the only possible clock event is for *phitwo* to fall again. This completes a full clock cycle. These events are shown in Figure 7.6.

Under these constraints, the clocked logic may be described by the following partial specification:

```
CCL(x,y) <= {din>p:noteq(p,x)}CCL(p,y)
 + {phione<true}{phione<false}CCL1(x,x,y)
CCL1(x,z,y) <= if noteq(comfun(z),y) then
      {phitwo<true,dout<comfun(z)}CCL2(x,y)
 + if eqs(comfun(z),y) then
      {phitwo<true}CCL2(x,y)
 + {din>p:noteq(p,x)}CCL1(p,z,y)
CCL2(x,y) <= {din>p:noteq(p,x)}CCL2(p,y)
 + {phitwo<false}CCL(x,y)
```

Explanation

The first two lines describe the initial state `CCL(x,y)` in which a pulse on *phione* may occur to latch the current value on *din* into an internal state. In state `CCL1(x,z,y)` a new output value, computed by the function `comfun(z)` whose argument is the latched internal value, may be read out by a pulse on *phitwo*. New values may be put on the input in this state without affecting the latched value. In the third state, a falling edge on *phitwo* is required to complete the two-phase clock cycle.

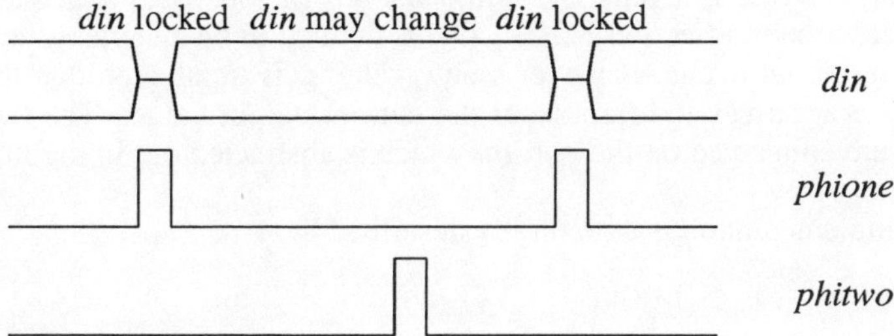

Figure 7.6: Timing diagram for 2-phase clock constraint

Now the implementation needs to be described. The latches have the following behaviour:

```
L1(x,y,false) <= if noteq(x,y) then
      {phione<true,lin<x}L1(x,x,true)
 + if eqs(x,y) then {phione<true}L1(x,x,true)
 + {din>p:noteq(x,p)}L1(p,y,false)
```

```
L1(x,y,true) <= {phione<false}L1(x,y,false)
 + {din>p:noteq(x,p)}L1(p,y,true)
L2 <= L1 [lout/din][dout/lin][phitwo/phione]
```

Explanation

The first set of states of L1 are those in which the value on the clock line is false. If it becomes true, the value on the input port is transferred to the output, provided of course that it is a different value from that already present. Alternatively, a new value may be placed on the input port, *din*. When the value on the clock line is true (as described in the 4th and 5th lines), the only possibilities are that it will become false or that a new value will be placed on the input. L2 has identical behaviour with a suitable renaming of ports as Figure 7.5 indicates.

The combinational logic can be described as follows:

```
LOGIC <= LOG * DEL2 - int
LOG(m,n) <=
    {lin>p:and(noteq(p,m),eqs(comfun(p),n))}
            LOG(x,n)
 + {lin>p:and(noteq(p,m),noteq(comfun(p),n)),
                    int<comfun(p)}LOG(x,n)
```

Explanation

This is an example of the constructive technique for specifying devices with delays. DEL2 is a generic 2 unit delay box of sort {*int, lout*}, such as was described in Section 3.2.3. LOG is defined to be delayless, producing an output event whenever a new value p is input such that the value of `comfun(p)` differs from the current output value. The two devices are connected on the port *int* which is abstracted out in the first line.

The implementation is structurally described by

```
IMP <=   L1 * LOGIC * L2 - lin - lout
```

To verify that this implementation satisfies the specification CCL, it needs to be shown that CCL * IMP can perform the same actions as CCL. The expansion of L1 * LOGIC * L2 - lin - lout is too lengthy to present here, having some 50 branches in the choice sum. Composition with CCL removes most of these branches, leaving only 16. In order to verify the implementation, the first step is to prove that these 16 branches can be matched with the 2 branches of CCL, using the rules for matching that were described in Section 5.2.4.

Now suppose that CCL had been specified fully rather than partially. With 3 inputs *din, phione* and *phitwo*, the full specification would need at least $2^3 - 1 = 7$ branches, and in all probability quite a few more than this. Since CCL would now be specified in such a way that all combinations of input events are possible, the 'pruning' effect of composing CCL with IMP would therefore have been much reduced. Both the specification and the implementation would, for example, contain branches whose guards involved events on *phione* and *phitwo* simultaneously. Thus the first step of proof, proving that the *initial* actions of the implementation and specification are equivalent, would now involve an attempt to match more than 7 branches with nearly 50, instead of matching 2 with 16 as was required above. Similarly, the task of establishing that subsequent actions of the specification and implementation are equivalent would be much more involved, because of the much wider range of input events that may take place (e.g. successive pulses on *phione*) and the greater number of states that may be reached. It is apparent, therefore, that by using a partial specification for a device, the task of verifying that an implementation meets that specification is greatly reduced in complexity.

The use of partial specifications coupled with constraints to ease the verification task has also been attempted in the HOL system [Herbert 88]. Device descriptions are split into three parts: the function, a set of input constraints, and a set of output stability assertions. The description of the function is essentially a partial specification; the device is guaranteed to perform this function if its input constraints are met, and this will result in certain stability conditions on the output (e.g. the output never changes during a certain part of the clock cycle). Verification of devices described in this way can then be simplified, as the functional behaviour is verified without regard for detailed timing phenomena; these are dealt with later, simply be ensuring that the output stability assertions of each device satisfy the input constraints of the devices to which they are connected. In effect, this work involves the separation of timing and function, as discussed in Section 5.2.4.

Constraints Arising From Verification

In the preceding Sections a number of reasons for using partial specifications have been proposed: they are easier to write than full specifications; partially specified devices may be easier to implement; and it may be easier to prove that an implementation satisfies a partial specification than a full one. It has been asserted throughout that a partial specification can only be legitimately used if a constraint is attached to it to ensure that events that are not accounted for in the partial specification are not generated by the environment. The benefits of partial specifications thus become the rationale for the use of constraints, even though the restrictions that they impose may require some additional effort in the design task, as was shown in Section 7.3.2.

The assertion that constraints are a necessary accompaniment to partial specifications has been based largely on intuitive arguments. The most powerful of these is the following: if a partially specified device is used in an environment that tries to perform an event which the partially specified device cannot accept, then the result may be a deadlock caused not by an incorrect implementation but simply by an inadequate specification. An example of this was first described in Section 3.2. By using a constraint in this situation, the behaviour of the environment could be checked before seeking to establish how the partially specified device would function in it; the fact that the environment did not satisfy the constraint could be detected, and either the environment would be redesigned or the device specified more fully.

There is a further argument for the need to attach constraints when partial specifications are used. This argument was developed by Traub [Traub 86] in his work on verification using CIRCAL and concerns the need to generate a constraint on the environment of a device that has been specified partially, designed and verified. In general, this device will form one of the boxes that implement another device at a higher level in the design hierarchy. Figure 7.7 shows a tree-structure representation of this hierarchy, in which the device currently being discussed is called box 'B'.

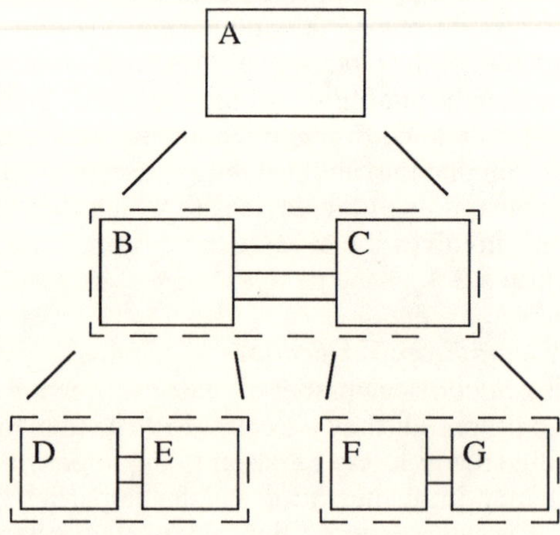

Figure 7.7: Boxes in a design hierarchy

Both B and C may have partial specifications. It will have been shown that these specifications interact in such a way as to implement the specified behaviour for A correctly. When B is designed, it will be shown that the behaviours of D and E interact in such a way as to satisfy the partial specification of B. Similarly, F and G will satisfy C. However, D * E will, in general, be able to perform actions additional to those of B; this is acceptable by the definition of satisfaction. Likewise, F * G will be able to perform more actions than C. These two sets of additional actions may interact in such a way that the combined behaviour of boxes B and C is no longer such that it correctly implements the specification A.

In order to deal with this problem, a contextual constraint can be defined which ensures that the additional, unwanted interactions cannot happen. Traub defines the *weakest safe context* as the 'least restrictive' contextual constraint that does this. The derivation of this is fairly involved, but the end result is that the weakest safe context is described by a device that rejects those actions that can be performed by D * E but not by B. Traub shows that if the environment of B satisfies this constraint in the sense defined in Section 7.3.1, then no unwanted interactions will occur. Thus the final reason for attaching a contextual constraint to a specification is to ensure that the implementation of the device will not interact unpredictably with the environment. Constraints that are attached for this reason should be dealt with in the design process using the techniques described in Section 7.3.2.

7.4   Constraints and SuperC

Throughout this Chapter, the uses of constraints have been illustrated with device and constraint specifications written in enhanced CIRCAL. However, SuperC has been proposed as a more 'friendly' interface to CIRCAL, so it is to be hoped that it can also support partial specification and the description of constraints. This Section discusses some extensions to SuperC to provide these capabilities.

One of the reasons for the development of SuperC was to prevent a designer from specifying devices whose inputs could not always accept new values. That is, SuperC enforced the writing of full specifications. In this way, the problems that arise from using partial specifications, which were illustrated in Section 5.2.2, could be totally avoided. However, it has been demonstrated in this Chapter that partial specifications can be safely used provided they are accompanied by a suitable environmental constraint; furthermore there are numerous reasons, in addition to the reduction of effort of writing specifications, for using them. Therefore, if SuperC is to be useful as a design language, it is important that some way of writing partial specifications be built into it.

Partial specifications differ from full specifications in that a partially specified device may not necessarily accept all possible valid input events at all times. This difference can be subdivided into two broad categories:

- there may be certain *times* at which a value on a particular input port is not allowed to change (e.g. setup time constraint); or
- there may be certain combinations of *values* that may not appear on a set of ports (e.g. restriction on RS flip flop inputs).

Thus, to enable specifications written in SuperC, which are full by default, to be made partial, it is necessary to have some way of describing these two types of restriction.

7.4.1   Temporal Restrictions

To describe the first class of restriction, a new keyword `stable` could be introduced. (A similar feature is found in VHDL [USAF 84].) This would enable the writing of descriptions in which values may not change at certain times, as illustrated by the following example, a D-type flip flop with a hold time of 3 time units.

Example
```
dff(data,clk) = q
     where
          q = event(clk,true):data
     with constraint
          event(clk,true) => stable(data,3)
     end
```

Explanation
The line beginning 'q =' describes the behaviour of the flip flop as if it were to be fully specified. A rising edge on the clock port will cause the value on the input port (*data*) to be transferred to the output. The keywords 'with constraint' introduce the restrictions that will turn this into a partial specification. The syntax for describing the restriction is of the general form

```
event(port_a, value) => stable(port_b, duration)
```

The interpretation of the restriction in this case is this: when a rising edge occurs on the clock port *clk*, the value on the port *data* must remain stable for 3 ticks.

Having found a way to describe the restriction on a specification, one needs some formal way of translating this into enhanced CIRCAL. A fairly elegant method is:

1. Translate the part of the description that precedes the constraint into a *full* specification, as described in Section 3.3.
2. Translate the constraint specification into CIRCAL.

3. Compose the constraint specification with the full specification, thus reducing it to a partial specification.

Ignoring for the moment the fine details of step 2, there is an important point to be made here. In contrast to CIRCAL, it is not difficult to write a full specification in SuperC. The motivation for producing a partial specification is to assist the design and validation tasks, or to describe a library part whose behaviour can only be guaranteed for a limited range of input stimuli. The partial specification is therefore not produced at the outset, but is generated by the constraining of a full specification. A full specification is constrained by being composed with an abstract device that will only allow certain patterns of input events, i.e. a contextual constraint as defined at the start of this Chapter. The composition process removes certain branches from the full specification, thus reducing it to a partial one. This procedure may be considered a particular instance of the constructive specification technique of Section 3.2.2, as it relies on the composition of two behavioural descriptions to produce the actual specification.

A significant point here is that, like SuperC, many other languages do not allow partial specifications. However, this technique of composing a full specification with a constraint is applicable to any language that has a formal means for constructing behaviours. For example, the technique could readily be applied to higher-order logic.

An attractive feature of this approach to partial specification is that the constraint and the specification are guaranteed to be correctly 'matched'. That is, the constraint that is used to generate the partial specification will ensure that the only actions that the environment can perform are the exact ones that the partially specified device can accept. This follows from the fact that the actions that are 'pruned' from the full specification by composition with the constraint are the same actions that the constraint prohibits the environment from performing. This is in contrast to the approach described in Section 7.3.1, in which a partial specification is written at the outset and a constraint is then added in the hope of preventing actions that the specified device cannot accept.

Step 2 of the above procedure, the translation of the SuperC representation of a constraint into an abstract device described in CIRCAL, can be achieved fairly easily. The device must be such that it accepts all possible events until the trigger event occurs. It should then disable the port named after the keyword `stable` for the specified length of time. If, as before, the constraint is described by

```
event(port_a,z) => stable(port_b,duration)
```

then the device which provides the appropriate constraint is as follows:

```
CON <= {t} CON + {t,port_a>x:noteq(x,z)}CON
  + {t,port_b>x} CON
  + {t,port_b>x,port_a>x:noteq(x,z)}CON
  + {t,port_a<z}CON1(duration)
  + {t,port_b>x,port_a<z}CON1(duration)
CON1(n) <= if noteq(n,1) then
      ({t}CON1(n-1) + {t,port_a>x}CON1(n-1))
  + if eqs(n,1) then ({t}CON + {t,port_a>x}CON)
```

Explanation

CON is the initial state in which any event is possible. When the value z is placed on *port_a*, the device moves into state CON1(duration). In this state, no events are possible on *port_b*, but the counter variable n is decremented at every tick. After a number of ticks equal to duration, the device returns to the state CON, thus enabling *port_b* again.

7.4.2 Port Value Restrictions

The second class of restrictions, those which relate to the values that sets of ports may hold, can be represented simply by predicates that are true if the values on the ports are acceptable. The following example illustrates this.

Example

A device that might be specified with a restriction on the values on its input ports is a bus, which may only be driven by one device at a time:

```
bus(ina,inb) = out
  where
    out = if eqs(inb,ZZ) then event(ina,x):x,
          if eqs(ina,ZZ) then event(inb,x):x
  with constraint
    or(eqs(inb,ZZ),eqs(ina,ZZ))
  end
```

Explanation

The variables ina and inb are of a type that includes a special value 'ZZ' which represents the high-impedance or non-driving state in tristate logic. The behaviour of the bus is simple: if the value on the input *inb* equals 'ZZ', then any change on the other input *ina* will be passed directly to the output. The symmetric behaviour when *ina* is set to 'ZZ' is described by the third line. The constraint statement implies that at least one of the inputs must be set to the high-impedance value at any time, thus ensuring that the bus is never driven by both inputs at once, a situation that would have unpredictable results.

The translation of this type of constraint specification into CIRCAL is a little more difficult than translation of the first type as it relates to static values on ports rather than events. What is required is a device that can accept any events on the constrained ports except those that will lead to states prohibited by the predicate. If the predicate is represented by $P(\underset{\sim}{x})$, with $\underset{\sim}{x}$ representing the vector of values on a set of constrained ports, and the ports themselves are denoted by M, then the device CON which enforces the constraint will have the following behaviour:

$$\text{CON}(\underset{\sim}{x}) \Leftarrow \sum_{g_k \subset M \wedge g_k \neq \phi} \left(\bigcup_{x_i \in g_k} \{x_i > z_i : P(\underset{\sim}{x}')\}\text{CON}(\underset{\sim}{x}') \right)$$

where $\underset{\sim}{x}'$ represents the new vector of values on the ports of the device after the listed events have occurred. The g_ks represent all the possible subsets of M. The behaviour of CON is therefore a choice sum of branches, each branch having a guard which corresponds to one subset of M. The events in each guard are represented by the inputting of a value which is qualified by a conditional, thereby ensuring that the new vector of port values will satisfy the constraining predicate P. As before, composition of this device with the full specification will yield a partial specification which incorporates the desired constraint.

7.5 Summary

This Chapter has discussed the concept of constraints and their use in design and validation. This concept is a familiar one to designers and many examples of constraints appear in real design situations. However, constraints usually only receive an informal treatment, and the word itself is rarely defined.

The intuitive definition of a constraint as something that imposes a restriction was proposed, and a number of examples that fit this definition were examined. This led to the conclusion that constraints could be divided into two types: restrictions on the behaviour of a device's environment, and restrictions on the choices open to a designer. The first type, being more amenable to formal treatment, were studied in this Chapter.

The formal use of constraints requires, first of all, that they can be formally specified. A mechanism for doing this is to represent a constraint by an abstract device that can perform any actions except those that the constraint prohibits. This type of device is easily described in CIRCAL, since it enables the description of devices that cannot always perform all possible actions.

In order to establish that constraints are of widespread usefulness, their role in each of the three main subtasks of the proposed methodology was examined. The concepts of full and partial specifications were defined and it was shown that the use of the latter, in which not all combinations of valid input events are possible at all times, could reduce the effort required to write specifications. The use of such specifications, however, is only justified if a contextual constraint is written to ensure that the environment of a partially specified device will only attempt to perform those actions that the device will accept. A formal method of establishing that an environment satisfies a constraint was presented.

Partial specifications can also assist in the design task, and a simple example was used to show that the unnecessary detail that may be included in a full specification could result in the rejection of a simple and acceptable solution to a design problem. There is also a price to be paid in the design task for the use of constraints — environments must be designed so that they satisfy any contextual constraints imposed on them. A method of modifying undesigned environments to meet constraints was presented. If parts of the environment have been designed using parts from a library, it may be necessary to choose a different library part, or not to use a library part at all. In this situation, contextual constraints can be seen to guide the designer by eliminating certain options. It is also possible that, in order to satisfy a constraint, an already-designed device may have to be re-designed, so that top-down design is replaced by iterative design.

In the verification task, partial specifications were again seen to be of assistance. The reduced behaviour of a partially specified device enables the process of demonstrating equivalence between specification and implementation to be simplified considerably. The need for contextual constraints to accompany partial specifications was again demonstrated, the motivation in this case being to prevent unexpected interactions between implementations whose behaviours are richer than those of their specifications.

Finally, a mechanism for expressing partial specifications in the high-level language SuperC was presented. In this language, partial specifications are represented as full specifications with restrictions; the restrictions may be on the times at which port values may change or on the values that may be placed on certain ports. It was shown that a full specification can be made partial by composing it with a constraint, this approach having the advantage that it also ensures that any environment that meets the constraint will perform actions that the partial specification can accept.

The main point made in this Chapter was that a formal treatment of constraints can provide assistance in *all* tasks of a design and validation methodology. Conversely, it is only through the adoption of a formal, language-based methodology that the assistance provided by constraints can be made available.

EIGHT

CONCLUDING REMARKS

8.1 Summary of Arguments

The main argument developed in the preceding chapters could be summarised as follows: that the most suitable medium for the representation
of hardware during the design process is a formal language, as this enables the use of a hierarchical, integrated approach to design and validation. The book has examined both the reasons behind this argument and
its consequences.

The main reasons for adopting an integrated, hierarchical approach to
design and validation are to improve the likelihood of correctness of a design and to minimise the cost of design errors by enabling their detection
at the earliest possible stage. Hierarchical validation can only be carried
out during the design process if behaviour can be formally represented
at all levels of hierarchy; the most suitable medium for this representation is a formally defined language, as it offers the necessary expressive
power and the ability to reason about behaviour in a rigorous way.

Hardware Description Languages

Having made the case for languages as a design medium, one of the first
issues to be considered is what the nature of such a design language
should be. This issue was examined first on an abstract level by considering the nature of hardware behaviour itself and how this might be
described in a useful way. Key aspects of hardware behaviour were identified as the interdependency of values on ports, the changing of values
(or events), and the timing and sequencing of the occurrence of events.
Based on these observations about behaviour, it was proposed that a behavioural Hardware Description Language (HDL) should have facilities
for the description of these aspects. An additional feature, the ability to
describe behaviour over a wide range of levels of abstraction, was also
proposed.

A more concrete view of hardware description was presented through
the examination of a number of HDLs that exhibit the necessary features

proposed above. These languages were seen to differ quite considerably in the additional features they provide and in the way in which these are provided. This reflects the differing motivations for the production of the languages and the range of environments in which they were developed. Only a handful of languages support formal reasoning about behaviour, a factor that becomes important when the issue of verification is addressed. One such language, CIRCAL, which has evolved from work on the modelling of concurrent processes and the application of this work to hardware, was selected for particularly close examination. For the majority of examples of the behavioural description of hardware, CIRCAL was used.

Subtasks

The proposed hierarchical design methodology consists of a number of subtasks which were identified as specification, design and validation. These three tasks were discussed in Chapters 3,4 and 5 respectively.

The task that must be performed first is specification. 'Specification' was taken to denote any form of behavioural description that does not imply how the specified device is or will be constructed. The ways in which the chosen language can help or hinder the specification task were presented. A language may help the designer in this task simply by facilitating the writing of concise descriptions, or by encouraging 'accurate' specifications. Two ways in which a specification should be accurate were identified: it should reflect the real intentions of the designer, and it should as far as possible describe realistic hardware. Both these goals are only partially attainable.

The development of a useable language which encourages accuracy was traced by examining CIRCAL and numerous enhancements of it. While displaying those features that were identified as essential for hardware description, CIRCAL was found to be lacking as a specification language when examined in the light of the above discussion. In particular, descriptions in CIRCAL were found to be quite cumbersome for even fairly simple devices such as logic gates, and unacceptably so for more complex devices. Thus the pure language was enhanced with a number of features, such as the ability to write descriptions that are parameterised over state variables, and a richer variety of events.

Certain techniques may be used with a language to increase its ease of use and accuracy. A number of such techniques were presented for enhanced CIRCAL. In order to ensure that these techniques are followed, a 'higher-level' description language (by analogy to high-level programming languages) was developed. This offered restricted access to the features of CIRCAL but improved the ease of writing and accuracy of the resultant specification.

The specification task is followed by design. This task was seen to consist of two parts, partitioning and description. Separation of the two parts is rather difficult, as description (associating a behaviour with a structural object) is often performed informally in the designer's mind at the same time as partitioning. Partitioning consists of splitting a box (which has a behavioural specification but no known internal structure) into a set of smaller, interconnected boxes. Description then consists of *formally* assigning behavioural descriptions to the boxes. This part of the task is thus actually specification — formalising the informal idea of behaviour that exists in the designer's mind.

The skill of the design task is in deciding into what smaller boxes a box will be partitioned. Some choices lead to implementations that are more efficient by some criteria; design decisions may be made that ease the verification task; some decisions may lead to incorrect implementations. Ways in which the designer may be assisted in reaching the goals of efficiency and correctness were presented. Of particular interest in the context of this book was the way in which the language-based approach to design can offer some of this assistance. It was seen that correctness-preserving transformations and design automation tools both rely on behavioural description languages. It was also demonstrated that a design language may assist a designer by restricting the options open to him, thus reducing his 'search space'.

The third task of the methodology, which completes a design step, is validation. This is the attempt to show that the behaviour of the implementation that was produced in the design task is, in some sense, equivalent to the specification. The two methods of approaching this task are simulation and verification. The former is the testing of an implementation's behaviour by calculating its response to a set of input stimuli. The latter is the proof by mathematical means that the implementation performs correctly with respect to its specification.

Examination of each method of validation revealed a number of requirements for the specification language. In particular, only simulation can be carried out without the specification language having the facility for the derivation of behaviours of constructed devices. The vast majority of HDLs are therefore suitable for simulation. Two methods for the simulation of expressions written in CIRCAL were presented. Some of the specification techniques that had been proposed previously were justified by demonstration of the undesirable consequences of ignoring them.

The main shortcoming of simulation was seen to be its failure to guarantee the correctness of an implementation. The second approach to validation, formal verification, seeks to provide such a guarantee. The first requirement for a language for verification is that the structural operators of the language have a formal interpretation that allows the behaviour of a constructed device to be mathematically established from the behavioural description of its component parts.

Verification using CIRCAL was examined and, as with simulation, it was found that specification techniques have an important part to play. It was shown that correct implementations could be rejected if some of the previously developed techniques were not used. A technique that enables specifications to be written without detailed knowledge of how they will be implemented was presented. It was shown that verification could be effectively carried out with such specifications in spite of the greater amount of information present in the implementation.

It was observed that CIRCAL really only addresses a part of the verification problem, that of showing that the ordering and timing of events is correct. The other part, the demonstration of 'functional' correctness, requires facilities for reasoning that are not supported by the CIRCAL framework, but could be provided by some sort of logical framework. A method was presented by which functional and temporal aspects of a proof could be separated, so that functional correctness may be more easily established without interference from details of timing.

Contextual Constraints

Having discussed the three main tasks of a design and validation methodology, a concept that can provide assistance in all three tasks was introduced. The concept of contextual constraints was defined as a restriction imposed by a device on the behaviour of its environment, i.e. the pieces of hardware to which it is connected. The use of such constraints provides further justification for a language-based approach to design, as the formal use of constraints requires that they be formally specified.

In order to support the specification of constraints, a language must be able to specify events that may not happen or situations that may not arise. CIRCAL was observed to be particularly suitable for this because it naturally allows the specification of a subset of all possible events, so that some events are disallowed.

It was demonstrated that contextual constraints, formally described, could provide assistance in each of the tasks of the proposed methodology. In specification, constraints can be used to ensure that the assumptions under which a specification is written are actually satisfied by its environment. In this way, *partial* specifications, which do not take account of all possible combinations of input events, can be safely used, thus reducing the difficulty of writing specifications.

The design task may also be assisted if partial specifications are used. The implementation of a device that is fully specified may be unnecessarily complicated by the need to respond correctly to input stimuli that will in fact never be applied. A contextual constraint is required in this instance to ensure that the environment generates only those input stimuli of which the specification takes account.

Finally the validation task may also be assisted by the use of partial specifications. The example of a system controlled by a two-phase clock showed that the magnitude of the task of proving equivalence between specification and implementation could be greatly reduced if the verification were carried out under the assumption that only selected patterns of stimuli would be applied to the system. A further use of constraints in verification was also discussed: to ensure that implementations do not interact in unpredictable and undesirable ways when they are capable of more actions than the specifications that they satisfy.

There is a *cost* of using constraints, which is that they must be satisfied by the environments to which they apply. This factor must be taken into account during the description phase of the design task. The description of a box must satisfy all constraints that are applied to it by parts to which it is connected. If the box is not already described when these constraints are generated, they can easily be incorporated into its description. If it has already been described, however, then it may be necessary to modify its description. This may mean that design decisions that were made at higher levels in the design hierarchy need to be revised, so that top-down design is replaced by iterative design.

8.2   Future Work

Having now studied in some depth the current state of the art in language-based design and verification, it is appropriate to consider briefly some potential avenues for future work in this field. A number of possibilities are described in the following sections.

8.2.1   Language Development

The most important theme in this book has been the role of languages in hierarchical design and validation. It has been shown that the features of a language strongly influence its usefulness in the various tasks and that the techniques of using the language are equally important. Thus, one of the most important fields for future work is that of language development. While some languages in the past have been developed in isolation, there are strong arguments for developing a language with the specific aim of supporting the integration of validation and design.

The development of the language SuperC suggests a possible route that could be followed for the development of a language to support a design and validation methodology. By giving the designer access to only certain features of CIRCAL the design process is assisted: the designer is prevented from falling into certain traps and the difficulty of the task of capturing his ideas formally is reduced, thus improving the chances that it will be done correctly. So, while SuperC may fall well short of the 'ideal design language', it represents a first step down a promising avenue of language development.

8.2.2 *Assistance for Design*

Whereas it is easily shown that a language-based methodology is essential for the validation of implementations, it has been less amply demonstrated that the language-based approach assists the design task. There are certainly some indications that it can: contextual constraints, which have been shown to assist the design task, can only be formally used in a language-based methodology; the idea of restricting a designer's access to the design language, which was seen to aid specification, may also be applied to design.

The development in the past of high-level programming languages was a considerable aid to the design of software systems; by enabling the programmer to work at more abstract levels using objects such as real numbers and lists rather than bits and addresses, high-level languages have greatly eased the task of software development and improved the chances of designing correct systems. It certainly seems reasonable to suppose that the development of suitable languages for hardware could lead to similar gains in the field of hardware design. The analogy between software and hardware could even be extended to include compilation; software compilers are now quite well understood, while compilers for hardware are still at a comparatively immature stage. It is to be hoped that future work on design languages and their use to direct the design process might lead to advances similar to those which have been made for software.

8.2.3 *Verification*

While the last few years have seen large advances in the field of hardware verification, there are still many problems to be overcome in making this process less complex and more amenable to machine assistance. It has been shown that CIRCAL provides considerable assistance in some aspects of the verification task, notably those that involve temporal and sequential, rather than functional, behaviour. In dealing with this latter aspect, theorem provers such as HOL [Gordon 85] and Boyer-Moore [Boyer 81] show promise. The development of tools of this kind may well lead to greater success in verification in the future.

In conclusion, the way forward for VLSI design is believed to lie with a formal, language-based approach. In this way, validation can be integrated with hierarchical design to minimise the chances and cost of design errors. The development of languages to support this kind of approach, and of closely related techniques for the use of such languages, should lead to greatly improved efficiency and correctness in VLSI design.

GLOSSARY

The following definitions should not be regarded as universal. They represent the definitions used in this book; many of the words are defined differently by other authors, although none of the definitions used here is totally at variance with accepted usage.

action A set of events which a device may perform.

behaviour The way in which the values on the ports of a device depend on each other and on the passage of time.

black box A device whose internal structure is not known or specified.

branch A part of a CIRCAL behavioural description, consisting of an **action** and of an **end state** which is reached after the performance of the action. A CIRCAL behaviour consists of a choice sum of branches.

composition The process of establishing the behaviour of a device constructed from several behaviourally described components.

constraint Any statement or rule which places a restriction on something.

context The set of devices which may communicate with another device constitute its context.

contextual constraint A restriction on the behaviour of a device's context, usually either in terms of the times at which events may occur or the values which may be placed on ports.

description Any statement that conveys information about a device. The information may relate to the device's behaviour, structure or geometry.

design step The complete process of producing and validating an implementation of a device at the next step down in the hierarchy.

device Any piece of hardware, not necessarily physically realisable.

domain One of the three possible types of description, either behavioural, structural or geometrical.

end state In a CIRCAL description, the behaviour which follows a guard.

enhanced CIRCAL   A language derived from CIRCAL in which states may be parameterised over variables of any type and events may involve the passing of values, variables or functions.

environment See **context**.

event An occurrence on a port. Usually this involves a change of the value which is present on the port, although in CIRCAL events may simply be synchronisation pulses in which no value is communicated.

guard A set of events.

implementation A representation of a device which contains information as to how it will be constructed, in terms of the interconnection of smaller devices whose behaviour is either known or specified.

levels (of abstraction) The layers in a design hierarchy at which more or less structural detail is available. At the top level, the only structural information is the ports of the device; at the next level the device is described as the interconnection of smaller devices; these smaller devices are described structurally at the next level down and so on.

partition To split a device into a collection of smaller devices and specify their interconnection.

port A channel through which a device may communicate with other devices. A port which may actually be realised in hardware is a *physical* port.

pure CIRCAL   The original CIRCAL calculus [Milne 83a] as presented in Chapter 2.

simulation The testing of the behaviour of a device by establishing its response to a (normally limited) set of input stimuli. If the response to all possible sets of input stimuli is tested, the simulation is termed *exhaustive*.

sort The set of ports through which a device may communicate with other devices.

specification A description of the behaviour of a device which does not directly imply anything about its structure. A *constructive* specification uses structural operators for descriptive purposes without committing the designer to constructing the device in the same way.

structure The way in which a device is constructed, described in terms of the interconnections of smaller devices.

synchronisation The occurrence of an event on a port which is shared between two or more devices.

synchronisation event A CIRCAL event which involves no passing of values, but a port name only. Also called a pulse.

tick An synchronisation event used to represent the passage of one unit of time.

validation Any attempt to establish that the behaviour of an implementation meets a specification, whether by simulation or formal, mathematical proof.

verification The attempt to prove formally that an implementation meets a specification under all (or most) conditions.

SUMMARY OF NOTATION

The following summary covers only those descriptive frameworks which are used more than once in the book.

B.1 *Predicate Logic*

The basic logical operators of predicate logic are presented below. These are used in both higher-order logic and temporal logic. In the latter case, temporal operators (not shown here) are also used. Note that P and Q are predicates and x is a parameter.

expression	meaning
$\equiv$	*is equivalent to*
$\neg P$	NOT P
$P \wedge Q$	P AND Q
$P \vee Q$	P OR Q
$\forall x.P$	P *is true for* all x
$\exists x.P$	P *is true for* some x
$P \supset Q$	P *implies* Q

B.2 *Pure* CIRCAL

In the following list, A and B are behaviours, and x and y are port names.

expression	operator name	meaning
`<=`	definition	
`A + B`	deterministic choice	The behaviour will be either A or B, depending on actions of the environment.
`A @ B`	nondeterministic choice	The behaviour will be either A or B, depending on actions unseen by the environment.
`{x}A`	guarding	After an event on x, the behaviour will be A.
`{x,y}A`	guarding	After simultaneous events on x and y, the behaviour will be A.
`/\`	deadlock	No actions possible.
`A * B`	composition	A is wired to B.
`A[x/y]`	relabelling	Port y is relabelled to x.
`A - x`	abstraction	Events on x are hidden from the environment and x is removed from the sort of A.

B.3 *Enhanced* CIRCAL

Events in enhanced CIRCAL may be more complex; the possible event types are tabulated below. a is a port, x and y are variables, p(x,y) is a predicate, and f(y) is a function. y is assumed to be a member of the set of current state parameters.

event type	example	meaning
pulse	`a`	As in pure CIRCAL.
input	`a>x`	A value is input on a and bound to x.
input (with predicate)	`a>x:p(x,y)`	A value is input on a and bound to x. x is such that p(x,y) is true.
output (value)	`a<3`	A fixed value (3) is output on a.
output (function)	`a<f(y)`	The value returned by f(y) is output on a.

GRAMMAR OF ENHANCED CIRCAL

The following is a specification, in `yacc` [Johnson 85] input format, of a context-free grammar for enhanced CIRCAL, allowing the passing of values of any type and the use of any user-defined function to specify conditionals, output values and state parameter assignments.

```
spec          : defn
              | spec defn
              ;

defn          : name params '<=' branches ';'
              ;

name          : STRING
              ;

params        :       /*empty*/
              | '(' paramlist ')'
              ;

paramlist     : par
              | paramlist ',' par
              ;

par           : STRING
              | INT
              | BOOL
              ;

branches      : branch
              | branches '+' branch
              ;
```

```
branch        : cond guard name assignments
              | cond '(' branches ')'
              ;

cond          :           /*empty*/
              | IF STRING THEN
              | IF function args THEN
              ;

function      : STRING
              ;

args          : arg
              | '(' arglist ')'
              ;

arglist       : arg
              | arglist ',' arg
              ;

arg           : STRING
              | INT
              | BOOL
              ;

guard         : '{' eventlist '}'
              ;

eventlist     : event
              | eventlist ',' event
              ;

event         : input
              | output
              | synch
              ;

input         : portname '>' arg incond
              ;

incond        :           /* empty */
              | ':' function args
```

```
                    ;

output          : portname '<' arg
                | portname '<' function
                ;

synch           : portname
                ;

portname        : STRING
                ;

assignments:      /* empty */
                | '(' asslist ')'
                ;

asslist         : assig
                | asslist ',' assig
                ;

assig           : function args
                | args
                ;
```

APPENDIX D

FUNCTION AND DATATYPE DEFINITIONS FOR COMPUTER SPECIFICATION

```
(* all integer types are defined as int, their
   sizes being taken care of by the functions
   which operate on them *)
type int2 = int;
type int13 = int;
type int16 = int;


(* a memory is represented as an ordered binary
   tree *)
datatype memory = empty
                | leaf of int13*int16
                | node of int13*memory*memory;


(* an enumerated type is used for opcodes *)
datatype opcode = HALT
                | JMP
                | JZRO
                | ADD
                | SUB
                | LD
                | ST
                | SKIP;


(* Define some auxiliary functions and
   constants *)
fun tworaise(n) =
       if n = 0 then 1 else 2*tworaise(n-1);
val maxint2 = 3;
val maxint13 = tworaise(13) -1;
val maxint16 = tworaise(16) -1;
val defval = 0;
```

```
(* Truncate a 16-bit integer to 13 bits *)
fun truncate(x) = x mod (maxint13 + 1);


(* Extract the opcode from a 16-bit integer*)
fun getop(x) =
   let val opbits = x div (maxint13 + 1) in
      if opbits = 0 then HALT else
      if opbits = 1 then JMP else
      if opbits = 2 then JZRO else
      if opbits = 3 then ADD else
      if opbits = 4 then SUB else
      if opbits = 5 then LD else
      if opbits = 6 then ST else SKIP end;


(* Store data at addr in memory *)
fun store(empty,data,addr) = leaf(data,addr)
  | store(m as leaf(a,d),data,addr) =
         if addr > a then
            node(a,m,leaf(addr,data))
          else if addr = a then leaf(a,data)
          else node(addr,leaf(addr,data),m)
  | store(node(ad,x,y),data,addr) =
       if addr > ad then
            node(ad,x,store(y,data,addr))
          else node(ad,store(x,data,addr),y);


(* Retrieve the data stored at addr *)
fun fetch(empty,addr) = defval
  | fetch(leaf(a,d),addr) =
         if a = addr then d else defval
  | fetch(node(ad,x,y),addr) =
          if addr > ad then fetch(y,addr)
           else fetch(x,addr);


(* Add two 16-bit integers. *)
fun add16(a,b) =
       let val sum = a + b in
          if sum > maxint16 then
            sum - maxint16 else sum end;


(* Subtract two 16-bit integers. *)
fun sub16(a,b) =
```

```
let val dif = a - b in
if dif < 0 then 0 else dif end;
```

BIBLIOGRAPHY

[Aylor 86]   James H. Aylor, Ron Waxman, and Charles Scarrat.
                    VHDL — feature description and analysis. *IEEE
                    Design and Test*, 3(4):17–27, April 1986.

[Babiker 83]   S. A. Babiker, R. A. Fleming, and R. E. Milne. A
                    tutorial for LTS. Technical Report ITM 224.83.3,
                    Standard Telecom Laboratories Ltd., 1983.

[Babiker 85]   S. A. Babiker, A. J. Evans, and Robert E. Milne.
                    A guide to designing with LTS. Technical Report
                    ITM 311.85.5, Standard Telecom Laboratories Ltd.,
                    1985.

[Backus 78]   J. Backus. Can programming be liberated from the
                    von Neumann style? *Communications of the ACM*,
                    21(8):613–641, August 1978.

[Boyer 81]   Robert S. Boyer and J. Strother Moore. *A Computa-
                    tional Logic*. ACM Monographs Series. Academic
                    Press, 1981.

[Bryant 81]   Randal E. Bryant. MOSSIM: A switch-level simu-
                    lator for MOS LSI. In *Proceedings of 18th Design
                    Automation Conference*, pages 786–790, 1981.

[Bryant 84]   R. E. Bryant. A switch-level model and simula-
                    tor for MOS digital systems. *IEEE Transactions on
                    Computers*, C-33(2):160–177, February 1984.

[Bryant 86]   Randal E. Bryant. Can a simulator verify a cir-
                    cuit? In *Formal Aspects of VLSI Design, Proceed-*

ings of Edinburgh Workshop on VLSI, pages 125 – 136, 1986.

[Burns 88]   S. M. Burns and A. J. Martin.   Synthesis of self-timed circuits by program transformation. In *Proceedings of IFIP WG 10.2 International Working Conference*, pages 97–114, Glasgow, 1988.

[Camilleri 86]   Albert Camilleri, Mike Gordon, and Tom Melham. Hardware verification using higher-order logic. In *Proceedings of IFIP WG 10.2 International Working Conference*, pages 41–66, Grenoble, 1986.

[Camurati 87]   Paulo Camurati and Paulo Prinetto. Formal verification of hardware correctness: An introduction. In M. Barbacci and C. J. Koomen, editors, *Proceedings of the 8th International Conference on Computer Hardware Description Languages and their Applications*, pages 225–247, Amsterdam, April 1987.

[Carlstedt 86]   T. Carlstedt-Duke. Experiences using EDIF. In *Proceedings of Third Silicon Design Conference*, pages 89–93, London, July 1986.

[Carter 79]   W. J. Carter, W. H. Joyner, and D. Brand. Symbolic simulation for correct machine design.   In *Proceedings of the 16th Design Automation Conference*, pages 280–286, San Diego, CA, June 1979. ACM/IEEE.

[Cohn 87]   Avra Cohn.   A proof of correctness of the Viper microprocessor: The first level.   Technical Report 104, Computer Laboratory, University of Cambridge, January 1987.

[Darringer 79]   J. A. Darringer. The application of program verification techniques to hardware verification.   In *Proceedings of the 16th Design Automation Conference*, pages 375–381, San Diego, CA, June 1979. ACM/IEEE.

[Davie 86a]   Bruce S. Davie. Hardware description languages: Some recent developments. Technical Report CSR-

198-86, University of Edinburgh, 1986.

[Davie 86b]   Bruce S. Davie and George J. Milne. The role of behaviour in VLSI design languages. In *Proceedings of IFIP WG 10.2 International Working Conference*, pages 1–18, Grenoble, 1986.

[Davie 88]   Bruce S. Davie and George J. Milne. Contextual constraints for design and verification. In *VLSI Specification, Verification and Synthesis*, pages 257–265. Kluwer Academic Publishers, 1988.

[Dewey 86]   Allen Dewey and Anthony Gadient. VHDL motivation. *IEEE Design and Test*, 3(4):12–16, April 1986.

[Eveking 85a]   Hans Eveking. The application of CHDLs to the abstract specification of hardware. In *Proceedings of the 7th International Symposium on Computer Hardware Description Languages and their Applications, CHDL'85*. North-Holland, 1985.

[Eveking 85b]   Hans Eveking. Formal verification of synchronous systems. In Milne and Subramanyam, editors, *Formal Aspects of VLSI Design*, pages 137–151, Edinburgh, 1985.

[German 85]   Steven M. German and Karl J. Lieberherr. Zeus: A language for expressing algorithms in hardware. *Computer*, 18(2):55–65, February 1985.

[Gilman 86]   Alfred S. Gilman. VHDL — the designer environment. *IEEE Design and Test*, 3(4):42–47, April 1986.

[Gordon 81a]   Mike Gordon. A model of register transfer systems with applications to microcode and VLSI correctness. Technical Report CSR-82-81, University of Edinburgh, 1981.

[Gordon 81b]   Mike Gordon. A very simple model of sequential behaviour of nMOS. In *Proceedings of VLSI '81*, Edinburgh, 1981.

[Gordon 85]   Mike Gordon. HOL: A machine oriented formulation of higher-order logic. Technical Report 68, University of Cambridge Computer Laboratory, 1985.

[Gordon 86]   Mike Gordon. Why higher-order logic is a good formalism for specifying and verifying hardware. In *Formal Aspects of VLSI Design, Proceedings of Edinburgh Workshop on VLSI*, pages 153–178, 1986.

[Hanna 85]   F.K. Hanna and N. Daeche. Specification and verification using higher-order logic. In *Proceedings of 7th International Symposium on Computer Hardware Description Languages and their Applications*, 1985.

[Harper 86]   Robert Harper, David MacQueen, and Robin Milner. Standard ML. Technical Report CSR-209-86, University of Edinburgh, 1986.

[Hatcher 82]   W. Hatcher. *The Logical Foundations of Mathematics*. Pergamon Press, 1982.

[Hayes 86]   J.P. Hayes. Digital simulation with multiple logic values. *IEEE Transactions on Computer Aided Design, CAD-5*, pages 274–283, April 1986.

[Herbert 88]   John Herbert. Temporal abstraction of digital designs. In G. Milne, editor, *Proceedings of IFIP WG 10.2 International Working Conference*, pages 1–25, Glasgow, July 1988.

[Hoare 78]   C. A. R. Hoare. Communicating sequential processes. *Communications of the ACM*, 21(8):666–677, August 1978.

[Hunt 86a]   Warren Hunt. *FM8501: A Verified Microprocessor*. PhD thesis, University of Texas at Austin, 1986. Report 47.

[Hunt 86b]   Warren A. Hunt. The mechanical verification of a microprocessor design. In *Proceedings of IFIP WG 10.2 International Working Conference*, pages 85–114, Grenoble, 1986.

[IEEE 87]   CAD Language Systems Inc. *VHDL Language Reference Manual, IEEE Draft Standard 1076B*, April 1987.

[Johnson 85]   S. C. Johnson. Yacc: Yet Another Compiler Compiler. Computer Science Technical Report No. 32, Murray Hill, 1975.

[Johnson 86]   Stuart G. Johnson. Graphical display of a concurrent device simulation using CIRCAL. Technical Report CSR-204-86, University of Edinburgh, 1986.

[Joyce 86]   Jeffrey Joyce, G. Birtwistle, and Mike Gordon. Proving a computer correct in higher-order logic. Technical Report 100, Computer Laboratory, University of Cambridge, December 1986.

[Kernighan 88]   Brain W. Kernighan and Dennis M. Ritchie. *The C Progamming Language, Second edition*. Prentice Hall, Englewood Cliffs, NJ, USA, 1988.

[Koomen 85]   Cees J. Koomen. The entropy of design: A study on the meaning of creativity. *IEEE Transactions on Systems, Man and Cybernetics*, SMC-15(1):16–30, January 1985.

[Langevin 88]   M. Langevin, C. Berthet, and E. Cerny. Verification of input constraints for synchronous circuits. In G. Milne, editor, *Proceedings of IFIP WG 10.2 International Working Conference*, pages 137–155, Glasgow, July 1988.

[Lattice 85]   Lattice Logic Ltd., 9 Wemyss Place, Edinburgh, U. K. *CHIPSMITH, a random logic compiler for gatearrays, optimised arrays and standard cells*, 1985.

[Lieberherr 83]   Karl J. Lieberherr and S. E. Knudson. Zeus: A hardware description language for VLSI. In *Proceedings of the 20th Design Automation Conference*, pages 27–29, Miami Beach, June 1983. ACM/IEEE.

[Lieberherr 84]   Karl J. Lieberherr. Towards a standard hardware description language. In *Proceedings of the 21st De-*

sign Automation Conference, pages 265–272, Alberquerque, NM, June 1984. ACM/IEEE.

[Lipsett 86]   Roger Lipsett, Erich Marschner, and Moe Shahdad. VHDL — the language. *IEEE Design and Test*, 3(4):28–41, April 1986.

[Lowenstein 86]   Al Lowenstein and Greg Winter. VHDL's impact on test. *IEEE Design and Test*, 3(4):48–53, April 1986.

[Luk 88]   Wayne Luk and Geraint Jones. From specifications to parameterised architectures. In *Proceedings of IFIP WG 10.2 International Working Conference*, pages 263–284, Glasgow, 1988.

[Martin 86]   Alain Martin. Self-timed FIFO: An exercise in compiling programs into VLSI circuits. In *Proceedings of IFIP WG 10.2 International Working Conference*, pages 115–135, Grenoble, 1986.

[Mead 80]   Carver Mead and Lynn Conway. *Introduction to VLSI systems*. Addison-Wesley, Reading, MA., 1980.

[Melham 87]   Thomas F. Melham. Abstraction Mechanisms for Hardware Verification. Technical Report 106, Computer Laboratory, University of Cambridge, May 1987.

[Milne 79]   George J. Milne and Robin Milner. Concurrent processes and their syntax. *Journal of the ACM*, 2(26):302–321, April 1979.

[Milne 80]   George J. Milne. The representation of communication and concurrency. Technical Report 4088, California Institute of Technology, September 1980.

[Milne 83a]   George J. Milne. CIRCAL: A calculus for circuit description. *Integration, the VLSI journal*, 1(2 & 3):121–160, October 1983.

[Milne 83b]   George J. Milne. The correctness of a simple silicon compiler. In *Proceedings of 6th International Symposium on Computer Hardware Description Languages and their Applications*, pages 1–12, Pittsburgh, 1983.

[Milne 84] George J. Milne. A model for hardware description and verification. In *Proceedings of 21st Design Automation Conference*. IEEE Computer Society Press, 1984.

[Milne 85a] George J. Milne. CIRCAL and the representation of communication, concurrency and time. *ACM Transactions on Programming Languages and Systems*, 7(2), April 1985.

[Milne 85b] George J. Milne. Simulation and verification: related techniques for hardware analysis. In *Proceedings of 7th International Symposium on Computer Hardware Description Languages and their Applications*, 1985.

[Milne 86a] George J. Milne. Towards verifiably correct VLSI design. In *Formal Aspects of VLSI Design, Proceedings of Edinburgh Workshop on VLSI*, 1986.

[Milne 86b] Robert E. Milne. Design transformation and chip planning. In *Formal Aspects of VLSI Design, Proceedings of Edinburgh Workshop on VLSI*, pages 23–43, 1986.

[Milne 88a] George J. Milne. The formal analysis of hardware timing. Technical Report HDV-2-88, University of Strathclyde, 1988.

[Milne 88b] George J. Milne and Mauro Pezze. Typed CIRCAL: A high-level framework for hardware verification. In G. Milne, editor, *Proceedings of IFIP WG 10.2 International Working Conference*, pages 115–136, Glasgow, July 1988.

[Milner 80] Robin Milner. *A Calculus of Communicating Systems*, volume 92 of *Lecture Notes in Computer Science*. Springer Verlag, 1980.

[Milner 83] Robin Milner. Calculi for synchrony and asynchrony. *Theoretical Computer Science*, 25(3):267–310, 1983.

[Morison 85] J. D. Morison, N. E. Peeling, and T. L. Thorp. The design rationale of ELLA, a hardware design and description language. In *Proceedings of 7th International Conference on Computer Hardware Description Languages*. North-Holland, 1985.

[Morison 86] J. D. Morison, N. E. Peeling, and T. L. Thorp. *The ELLA Language Reference Manual*. Praxis Systems plc, Bath, England, 1986. Issue 2.0.

[Moszkowski 83] B. Moszkowski. A temporal logic for multi-level reasoning about hardware. In *Proceedings of 6th International Symposium on Computer Hardware Description Languages and their Applications*, Pittsburgh, 1983.

[Moszkowski 85] B. Moszkowski. A temporal logic for multi-level reasoning about hardware. *Computer*, 18(2):10–19, February 1985.

[Nagel 75] L. W. Nagel. Spice2: A computer program to simulate semiconductor circuits. Technical Report ERL-M250, University of California, Berkeley, 1975.

[Nash 84] J. D. Nash. Bibliography of hardware description languages. *A.C.M. SIGDA Newsletter*, 14(1):18–37, February 1984.

[Pezze 87] Mauro Pezze. Behavioural abstraction and circuit verification using CIRCAL. Technical Report CSR-251-87, University of Edinburgh, 1987.

[Piloty 82] R. Piloty and D. Borrione. The CONLAN project: Status and future plans. In *Proceedings of 19th Design Automation Conference*, pages 202–212, Las Vegas, 1982. ACM/IEEE.

[Piloty 85] R. Piloty and D. Borrione. The CONLAN project: Concepts, implementations, applications. *Com-

puter, 18(2):81–93, February 1985.

[Sanella 85]   Donald Sanella and Andrzej Tarlecki. Program specification and development in standard ml. In *12th Symposium on Principles of Progamming Languages*, New Orleans, January 1985. ACM.

[SCS 87a]   Silicon Compiler Systems, 2045 Hamilton Ave., San Hose, CA, USA. *GDT Database and Language Tools Reference, Version 3.0*, June 1987.

[SCS 87b]   Silicon Compiler Systems, 2045 Hamilton Ave., San Hose, CA, USA. *Lsim Mixed-mode Analog and Digital Simulator Reference, Version 3.0*, June 1987.

[Shahdad 85]   M. Shahdad. VHSIC hardware description language. *Computer*, 18(2):94–103, February 1985.

[Sheeran 83]   Mary Sheeran. *μFP, an Algebraic VLSI Design Language*. Ph.D. thesis, University of Oxford, November 1983.

[Sheeran 86]   Mary Sheeran. Describing and reasoning about circuits using relations. In *Proceedings of Leeds Workshop on Theoretical Aspects of VLSI Design*, 1986.

[Sheeran 88]   Mary Sheeran. Retiming and slowdown in ruby. In G. Milne, editor, *Proceedings of IFIP WG 10.2 International Working Conference*, pages 285–304, Glasgow, July 1988.

[Siskind 82]   J. Siskind, J. Southard, and K. Crouch. Generating custom high-performance VLSI designs from succinct algorithmic descriptions. In *Proceedings of MIT Conference on Advanced Research in VLSI*, 1982.

[Subramanyam 88]   P. A. Subramanyam. Contextual constraints, temporal abstraction and observational equivalence in VLSI design. In G. Milne, editor, *Proceedings of IFIP WG 10.2 International Working Conference*, pages 156–182, Glasgow, July 1988.

[Terman 83]   C. Terman. *Simulation Tools for Digital LSI Design*. Ph.D. thesis, MIT, 1983. MIT/LCS/TR-304.

[Traub 86]   Niklaus G. Traub. *A Formal Approach to Hardware Analysis*. Ph.D. thesis, University of Edinburgh, 1986.

[USAF 84]   U.S. Air Force, Air Force Systems Command, Aeronautical Systems Division, Wright Patterson AFB, Ohio 45433, USA. *VHDL users manual*, 1984.

[Vaidya 83a]   A. K. Vaidya, D. L. Dietmyer, and M. K. Engh. WISLAN — a CONLAN member for gate array design. In Uehara and Barbacci, editors, *Proceedings 6th International Symposium on Computer Hardware Description Languages and Their Applications*, pages 31–42, Pittsburgh, 1983.

[Vaidya 83b]   A. K. Vaidya, D. L. Dietmyer, and M. K. Engh. WISLAN — technology transformation and optimization. In Uehara and Barbacci, editors, *Proceedings 6th International Symposium on Computer Hardware Description Languages and Their Applications*, pages 43–54, Pittsburgh, 1983.

[Wirth 82]   N. Wirth. *Programming in Modula-2*. Springer-Verlag, New York, USA, 1982.